PRODUCING ROYAL JELLY

A Guide For The Commercial and Hobbyist Beekeeper

R.F. van Toor

Northern Bee Books

Producing Royal Jelly

© 2013 R.F. van Toor

This reprint of the 2006 Revised Edition Published in 2013 by

Northern Bee Books
Scout Bottom Farm
Mytholmroyd
Hebden Bridge HX7 5JS (UK)

ISBN 978-1-908904-26-3

Printed by Lightning Source, UK

Contents

List of Figures

List of Tables

List of Plates

1. Introduction

Royal jelly is a natural substance produced from glands of worker honey bees by the fermentation of pollen and honey. The jelly is fed by worker bees to larvae which have hatched from fertilised eggs. As a consequence, the worker larvae develop into queens. Royal jelly is also fed to worker adults as an enriched food source.

Among other components beneficial to human health, royal jelly contains the eight essential amino acids, the full vitamin B complex, acetylcholine (a powerful neurostimulant), testosterone, insulin-like peptides and an antibacterial and antibiotic component.

As people become more aware of the perceived therapeutic benefits from royal jelly, the world demand for this natural bee-derived product is increasing. Consumption in Japan, for example, rose from negligible amounts in 1959 to 140 tonnes in 1984 and to about 300 tonnes in 1995.

While commercial beekeepers and the health food industry have largely kept pace with, and satisfied, consumer demand for royal jelly in its various formulations, there are a large number of hobbyist beekeepers who are interested in producing royal jelly for their own consumption and for their friends.

This book pools available information on royal jelly. It provides the reader with detailed instructions on methods of manipulating the honey bee *Apis mellifera* to produce the fresh product for personal consumption, or for sale to wholesaler-distributors.

2.   Why Do Bees Produce Royal Jelly?

The mechanisms for differentiation of the three castes of the honey bee; the queen, worker (female) and drone (male), (Plate 1) vary, with pure royal jelly used only for production of queens. Drones, which perform no duties except to mate with queens, are produced when the queen lays an unfertilised egg (containing only one set of 16 chromosomes per body cell) into the large 6 mm diameter cells on the honeycomb. Worker jelly is fed to the drone during its development.

Both workers and queens can potentially develop from the same fertilised eggs, which have usually been laid by the queen into 5 mm diameter worker cells. These eggs contain 32 chromosomes per body cell. Whether the egg develops into one or other caste depends on cell shape, nutritional components and amount of the food fed by the workers, and the amount of secretions from the mandibular glands of the developing larvae.

When fertilised eggs are either laid in queen cells (Plate 2) by the queen or they are transferred by workers from worker cells to queen cells, they are destined to become queens. When the egg hatches, usually three days after being laid, workers are stimulated to feed the newly-hatched larva royal jelly. The hypopharyngeal and mandibular glands from which the main components of royal jelly are formed, are located in the head of worker bees (Figure 1).

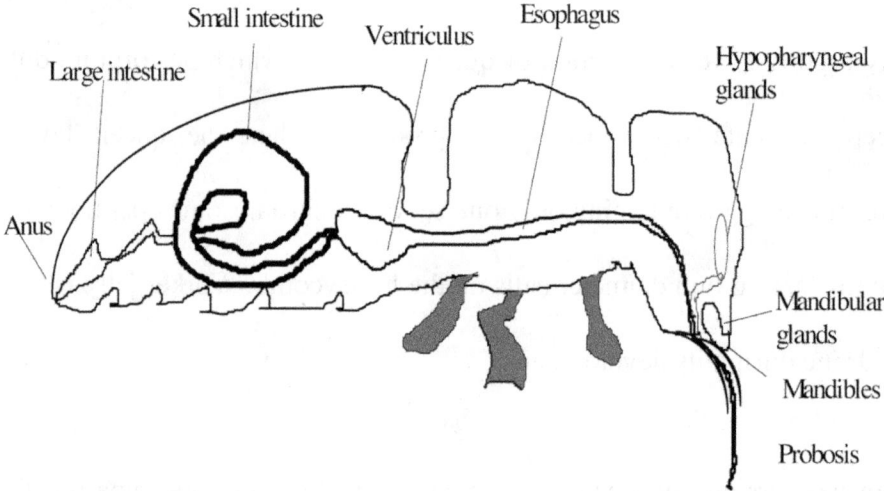

Figure 1: Schematic diagram of the alimentary canal of a worker bee.

The jelly fed to queen larvae differs from worker food in containing more mandibular gland secretions and in the amount fed. Queen larvae receive mostly mandibular and hypopharyngeal secretions during the first three days of feeding, and a 1:1 ratio of mandibular and hypopharyngeal secretions during the last two days of feeding. The mandibular secretions contain high levels of biopterin and pantothenic acid, giving 18 and 10 times higher concentrations than in worker food, respectively. The sugar content is also important and is the principal feeding stimulant for queen larvae. Queens are fed food containing 34% sugar, mainly glucose, in the first three days. On the critical third day of larval development, the high feeding rate and concentration of sugars stimulate the stretch receptors of the midgut in the larvae to release juvenile hormone from the corpora allata, a large globular organ found on the sides of the oesophagus. This release of juvenile hormone results in the formation of a queen. Glucose is also the main sugar fed during the remaining two days of the larval development.

The larvae develop for five days in the queen cell before the cell is capped. They are fed by nurse bees which tend to be 6-18 day-old worker bees. The nurse bees make about 150 visits to the larva before the cell is capped. The larva continues its development through the pre-pupa and pupa stages to emerge as an adult queen 16 days after egg laying (Table 1).

Table 1: Development stages of the queen bee.

Growth stage	Days after egg laying															
	1	2	3	4	5	6	7	8	9	10	11	12	13	14	15	16
Egg																
Hatching																
Larva 1st moult																
Larva 2nd moult																
Larva 3rd moult																
Larva 4th moult																
Larva 5th moult (pre-pupa)																
Pupae																
Adult 6th moult (emerging)																

In contrast, the food of worker larvae contains a ratio of 2:9:3 of mandibular gland secretions, hypopharyngeal gland secretions and pollen respectively, averaged over the first five days of larval feeding. The amount of food is much less than that fed to queens, and it has a lower concentration of sugar; 12% in the first three days of feeding. The sugar concentration increases to 47% with the addition of honey, with glucose being the predominant sugar in the early larval stages, and fructose the main sugar component in the later larval stages. Worker larvae are fed more hypopharyngeal gland secretions during the first few days of the larval stage than queens, and more honey and pollen during the last few days. This results in lower levels of juvenile hormone on days 3-5 than that for queen larvae, and consequently, the differentiation into workers. The worker adult emerges 21 days after egg laying, taking 14 days to develop through the pre-pupae and pupae stages instead of the nine days for the queen larvae. The development stages are shown in Plate 3.

Royal jelly is consumed not only by larvae but also by adult bees, including forages. Nurse bees transfer up to half their royal jelly to adult members of the colony, with younger workers receiving larger amounts of the jelly than older ones.

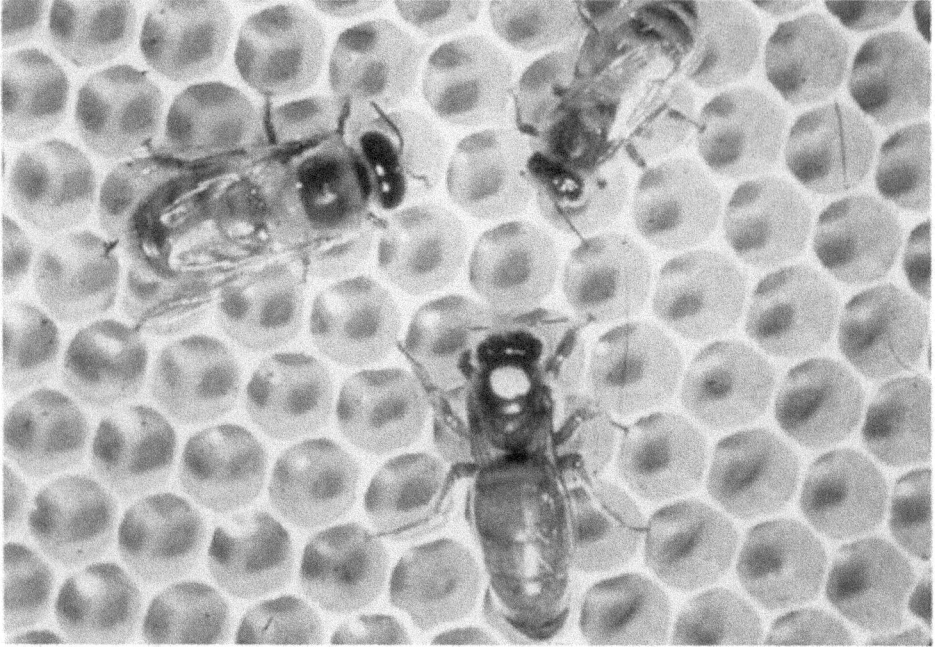

Plate 1: Healthy adult drone (left), worker (top right) and queen (bottom centre).

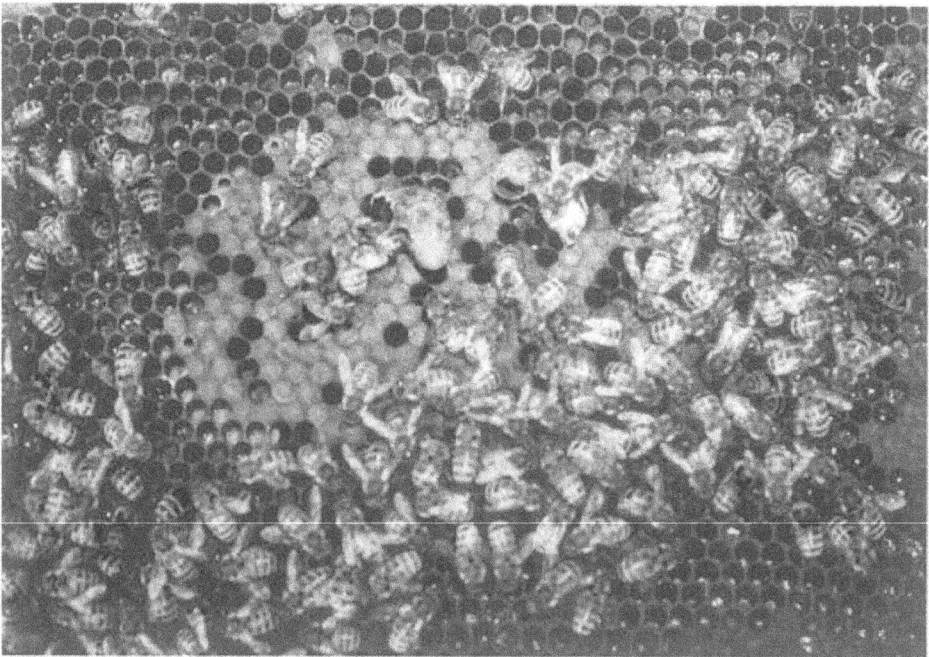

Plate 2: Formation of natural queen cells.

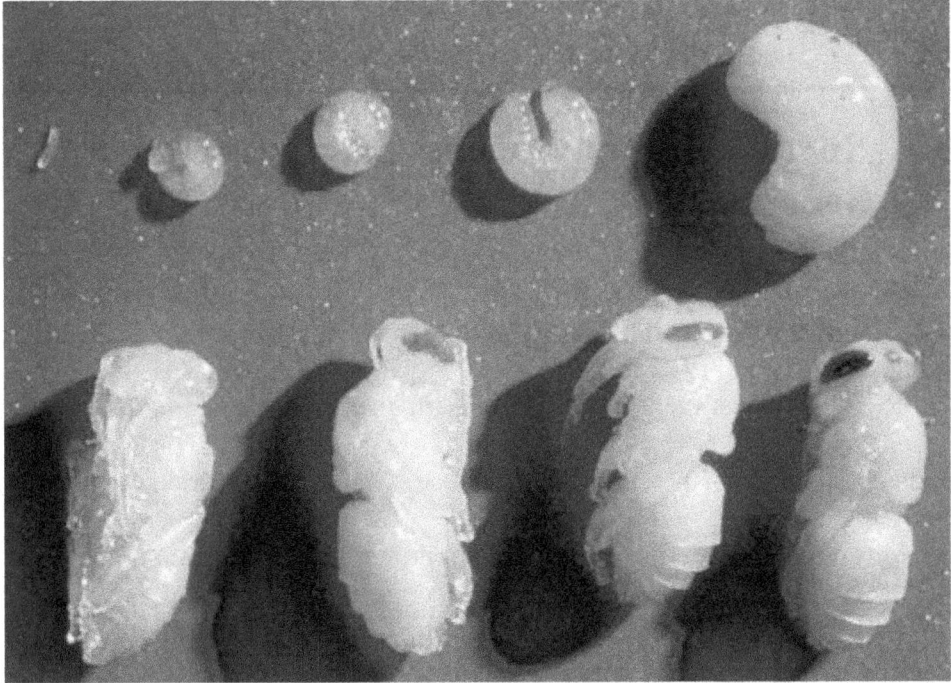

Plate 3: An egg (top left), and healthy worker larvae (top row) and pupae (bottom row) at different development stage.

3. Royal Jelly Components and Perceived Benefits

3.1 Chemical composition

Table 2 lists constituents that are present in significant amounts, or are considered to be of potential significance. About 3% of dried royal jelly remains unidentified, its properties unknown. Royal jelly contains two-thirds water, with the remainder rich in sugars, protein and lipids. It contains a large range of the vitamin B complex, although it lacks vitamins A, C and E.

3.2 Chemical Properties

Biological effects of royal jelly from specific components on animals and humans are diverse. 10-hydroxy-2-decenoic acid and a protein, royalisin, are unique to royal jelly. They are both known to have antibacterial activity, with the former also having antifungal activity. Royalisin kills bacteria because of its ability to disintegrate bacterial membranes. The antibiotic components have properties similar to gramicidin and penicillin.

Other effects that are identified to specific components include insulin which has glycaemic control to alter blood sugar levels. Testosterone is a male hormone. Acetylcholine is a powerful neurostimulant having vasodilator activity, and in royal jelly is thought to be responsible for an immediate feeling of euphoria. Glycoprotein promotes interferon activity and the lipid fraction reduces hypertension. Effects not identified with known components in royal jelly include reduction in tumour formation (Sarcoma-180 and Ehrlich ascites tumours), and inhibition of pathogenic bacteria.

Royal jelly is also rich in pantothenic acid, a part of coenzyme A. It is involved in metabolism of carbohydrates, fats and amino acids. Large amounts of fresh royal jelly consumed every day, e.g. 10 g, could make a significant contribution to the dietary intake of this vitamin.

Table 2: Composition of royal jelly per 100 g fresh material.

Component (units)	Average
Moisture (g)	66.8
pH	3.8
Total carbohydrates (g)	11.6
Glucose (g)	4.6
Fructose (g)	4.5
Sucrose (g)	1.1
Trisaccharides (g)	0.3
Total protein (g)	12.3
Free amino acids (g) [isoleucine, leucine, lysine, methionine, phenylalanine, threonine, tryptophan, valine]	0.25
Lipids (g)	5.1
10-hydroxy-2-decenoic acid (g)	2.5
10-hydroxydecenoic acid (g)	1.2
Ash (g)	1.0
Potassium (mg)	491
Sodium (mg)	37
Magnesium (mg)	30
Calcium (mg)	26
Zinc (mg)	2.7
Vitamin B complex	
B1 Thiamine (mg)	0.4
B2 Riboflavin (mg)	1.4
B3 Niacin (mg)	4.3
B5 Pantothenic acid (mg)	13.2
B6 Pyridoxine (mg)	0.5
B7 Mesoinositol (mg)	10.0
B8 Biotin (mg)	0.2
B9 Folic acid (mcg)	30
Minor components	
Benzoic acid (mg)	1.0
Acetylcholine (mg)	10
Testosterone (mg)	22
Insulin	Trace

Abridged from Lakin (1993).

3.3 Therapeutic uses

3.3.1 Animal studies

Many animal experiments studying the effects of royal jelly can be criticised for inadequate design, excessive levels of royal jelly, lack of precision and observation of nutritional effects rather than pharmacological effects. However, published studies strongly suggest that royal jelly can influence both the growth and development of animals.

Examples of favourable responses with oral administration of royal jelly to rats include:

- increased oxygen uptake and therefore activity by liver mitochondria (cell power houses) may support the use of royal jelly in the treatment of weak and debilitated patients,

- hypertension reduced by lipid (fat) fraction,

- antiflammatory effects and improved wound healing,

- increased body weight in males,

- exhibiting immunomodulatory properties leading to myeloprotection and antitumour activity

- increased production of serum luteinizing hormone, testosterone and progesterone in males,

- reduced levels of an indicator of oxidative damage (8-hydroxy-2-deoxyguanosine) in kidney DNA and serum implicated in extending the average life expectancy by 25% compared to the control group.

- increased thyroid and adrenal hormones in males, and

- increased survival following exposure to radiation.

In male hamsters, beneficial effects include:

- increased intra-testicular free testosterone and more intensive spermatogenesis from being fed 500 g/kg body weight for 12 weeks, with authors concluding that the long-term feeding of royal jelly inhibits the age-associated decline in the testicular function of male hamsters.

In rabbits, beneficial effects include:

- reduction of blood cholesterol levels induced by cholesterol-rich diets, and

- improved reproductive characteristics of females administered during rearing.

In pigs:

- freeze-dried royal jelly added to mixed feeds at 30 and 50 ppm improved weight gain in pigs (14.00% and 17.00%, respectively), feed utilization (8.00% and 13.00%), carcass yield (7.00% and 13.00%), meat yield (18.00% and 30.00%), and meat digestibility and tenderness.

In ewes:

- significant increase in lambing percentage.

And in calves:

- increased levels of red blood cells and gamma-globulin, and

- increased growth rates and resistance to infection.

3.3.2 Benefits in humans

Published studies on humans are one-off case histories rather than carefully monitored clinical trials. Claims for beneficial effects include those from patients suffering from fatigue, headaches, old age infirmity, tuberculosis and infectious hepatitis sufferers, anorexia, failure of children to thrive, inadequate lactation, asthma, and raised levels of blood triglycerides and cholesterol.

The amount of essential amino acids, B vitamins, essential fatty acids and trace elements consumed at the recommended dose rate of 500 mg/day is too small to have a nutritive effect on humans. Reports of the successful treatment of seriously debilitated patients through the use of much larger doses of royal jelly could have been due to nutritional intervention rather than any pharmacological effect.

However, some pharmacological claims are plausible. Antimicrobial and possibly hormonal effects can be attributed to known components of royal jelly. A hypotensive effect of reducing high blood pressure is claimed to be caused by the lipid component. Recently, a mechanism to explain the physiological effects from royal jelly in perceived improvement of menopausal symptoms has been proposed. Evidence was found in rats that royal jelly has estrogenic activities through interaction with estrogen receptors followed by endogenous gene expressions.

From published information and from numerous anecdotal reports, the commonest therapeutic use of royal jelly at the recommended dose is to alleviate debility (feebleness) and fatigue in otherwise healthy people or in people associated with a specific illness. Royal jelly can also be employed in conjunction with prescribed treatments to provide support and reinforcement to them, but not as an alternative or stand-alone treatment.

More thorough research is required before the definitive therapeutic benefits of royal jelly are known. People suffering from asthma or allergies are warned to take royal jelly with care. The proteins and pollens in the jelly may contribute to potentially fatal allergic reactions in isolated instances. Part of the allergic reaction in asthma sufferers can be a severe asthma attack which can be very difficult to reverse.

3.4 The recommended dose

The recommended dose of royal jelly for use as a dietary supplement for an adult is 500 mg fresh jelly per day. For children it is 250 mg per day. The adult amount of unprocessed royal jelly is roughly equivalent to a blob of jelly the size of the nail of the little finger. Yves Donadieu, a European doctor, recommends royal jelly should be taken over a period of four to six week course, with repeated courses as and when desired.

To obtain the best effects royal jelly it should be taken sublingually, that is, by slowly dissolving it under the tongue. This enables many of the active materials to be absorbed directly into the blood steam. Swallowing royal jelly is the next best method.

Munstedt, K. and von Georgi, R. (2003) in the American Bee Journal 143 (8), pages 647-650, summarises the scientific evidence for the health benefits of royal jelly. The authors cover mineral and vitamin composition in relation to the daily recommended dose. They also review preclinical studies on its biological activity and its safety, with specific emphasis on adverse reactions to royal jelly.

4.   Preparation For Production

4.1 Principals of production

The process of royal jelly production involves setting up a hive so that a portion of it is queen-less. Workers are stimulated by the reduction in queen pheromones to raise new queens. By placing plastic queen cells containing newly-hatched larvae in that portion of the hive, worker bees are stimulated to deposit royal jelly in to the cells to feed the developing queen larvae. Three days after larval transfer, the maximum amount of royal jelly which is surplus to the feeding larvae, is present in the cells. The larvae can then be discarded. The jelly is harvested by spatula or suction apparatus. Then, another batch of newly-hatched larvae can be transferred from brood frames directly into the queen cells to repeat the cycle.

4.2 Site selection

A sheltered site, centrally located to reduce travelling, is required. The site should provide early and regular supplies of natural pollen and nectar. A shed or hay barn (Plate 4) near the apiary is desirable for transfer of newly-hatched eggs to the queen cells e.g. by grafting, although grafting can be conducted comfortably from a vehicle. Shelter is also useful in making production comfortable in all weather, but again it is not essential.

4.3 Apiary structure

There are three types of hives used in royal jelly production. Those configured for:

1. royal jelly production,

2. standard honey hives called surrogate hives, and

3. nucleus hives called nucs.

Surrogate hives are used to provide extra pollen, honey, young and capped brood, and adult bees to boost ailing royal jelly producing hives. They are not necessary if production with strong bee colonies continues for less than six days (one or two cycles). But if production continues for more than nine days (three cycles), surrogate hives are necessary. This is because working royal jelly producing colony for this duration has the potential to deplete its resources. The surrogate hives should ideally be located about 100 m away from the royal jelly producing hives. This limits the possibility of any diseases present being transferred between sites by the promiscuous drones or robbing bees. Employ one surrogate hive for every two royal jelly producing hives (Table 3).

Nucs provide a convenient source of newly-hatched larvae. For convenience, they should be located with the royal jelly producing hives. One nuc is usually sufficient to supply enough newly-hatched larvae for three royal jelly producing hives.

Table 3: Recommended ratio of auxiliary hives to royal jelly producing hives.

Hive type	Number
Royal jelly production	6
Surrogate support hives	3
Nucs	2

Plate 4: An ideal site for royal jelly production.

Plate 5: A full-depth frame covered by approximately 3,500 bees.

4.4 Hive layouts

Royal jelly producing hives need to be positioned in a defined layout and painted in various colours to reduce drift. Drifting occurs when flying bees become disorientated and return to the wrong hive. Under severe drifting, some hives become over crowded and others become weak. The detrimental effects of drifting are to:

- Increase the risk of diseased bees, or foraging bees contaminated with insecticide, affecting otherwise healthy colonies.

- Increase the risk of spreading diseases, particularly American foul brood (Page 79) throughout the apiary as beekeepers swap brood and stores to even up the size of the colonies.

- Negate the selection of highest yielding hives based on genetic composition and other inherent characteristics of the colony.

4.4.1 Reducing drift

Drift can be reduced to acceptable levels by two means:

1. **Hive position:** Hives positioned in a straight line induce the highest incidence of drift. Drift is reduced if the hives are arranged in

squares with the entrances facing in all four directions, in a 'U' or circle with the entrances facing outwards, or in waves with the entrances facing in one broad direction. At sheltered sites, the direction of the hive entrance has little effect on royal jelly yields, although at exposed sites, hives with entrances facing the prevailing weather and away from the sun may show reduced performance.

2. **Colour:** Single coloured hives induce the most drift. Painting boxes different colours is effective in reducing drift. Bees cannot distinguish different shades of red, but can distinguish even slight differences in shades of blue. The colours most distinctive to bees are blue-green, yellow and black.

4.4.2 Hive set-up for reducing drift

To reduce drift, I recommend placing the hives in a 'U' shape, with the entrances facing outwards. This arrangement allows for a vehicle to be driven into the 'U' so that the vehicle is about the same distance away from all hives. Set each hive 1.5 m apart to allow ease of movement between hives. Paint the hive boxes different shades of blue and yellow.

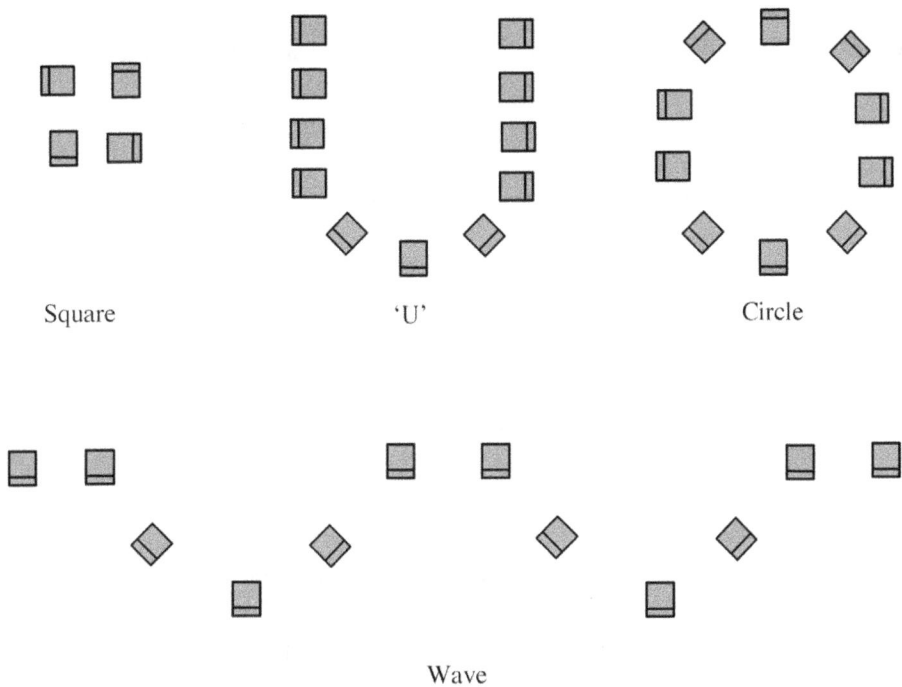

Figure 2: Hive layout to reduce drifting (double lines represent the entrance to a hive).

4.5 Hive design for royal jelly production

The hive should be arranged as in Figure 3. The queen is confined to the bottom box by a queen excluder, with a smaller 390 x 310 mm hardboard sheet placed on top and in the middle of the queen excluder. This sheet reduces the amount of pheromones produced by the queen from inhibiting royal jelly production in the second box.

Worker bees are still able to migrate back and forth through the slots in the queen excluder to the second box. The bottom box should contain a strong working colony. It consists of two frames of honey, two frames of pollen, two brood frames, four empty frames for the queen to lay in, and a one-year old queen. In the second box, there are two frames of honey, two frames of pollen, and four frames of brood which also include honey and pollen stores. One frame of honey can be replaced with a frame-type sugar feeder if an additional sugar source for the bees is required (Page 40). The brood frames in the second box should be a mix of capped and uncapped brood. The uncapped brood attracts the nurse bees close to the queen cells. Capped cells contain pre-emerged adults, which will eventually emerge near the queen larvae. Within six days after emerging, these young adults also become nurse bees. And one of the main duties of the nurse bees is to feed the queen larvae, which in turn increases the amount of royal jelly deposited into the queen cells. The frames holding the queen cell bars (Figure 4) are positioned in the middle of the second box.

The top (third) box becomes the honey super. A 10 x 10-20 mm front-facing entrance immediately above the second box allows bees returning with honey to enter directly into the nine-frame honey supers on top of the hive, or service the developing queens below the entrance. The size of this entrance should not exceed these dimensions because the decrease in air temperature in the hive caused by an increased draught from a larger

entrance will inhibit honey ripening and royal jelly production by the bees.

Box 3: Honey super with
syrup feeders if necessary

Honey entrance ————

—— Rim from a queen
excluder with the excluder
removed, and a 10 x 10-20
mm slot cut in the front.

Box 2: 2 frames queen cells
2-4 frames brood
2 frames pollen
2-4 frames honey

—— Queen excluder, with a
390 x 310 mm hardboard
pheromone excluder on
top.

Box 1: Active queen,
brood, pollen and honey

Main entrance ————

SIDE ELEVATION

Second box
(queenless)

1. Honey or sugar feeder
2. Pollen
3. Capped brood
4. Queen cells
5. Pollen & brood
6. Queen cells
7. Young brood
8. Capped brood
9. Pollen
10. Honey

First (bottom) box
containing queen

1. Honey
2. Pollen
3. Empty
4. Empty
5. Young brood
6. Young brood
7. Empty
8. Empty
9. Pollen
10. Honey

PLAN

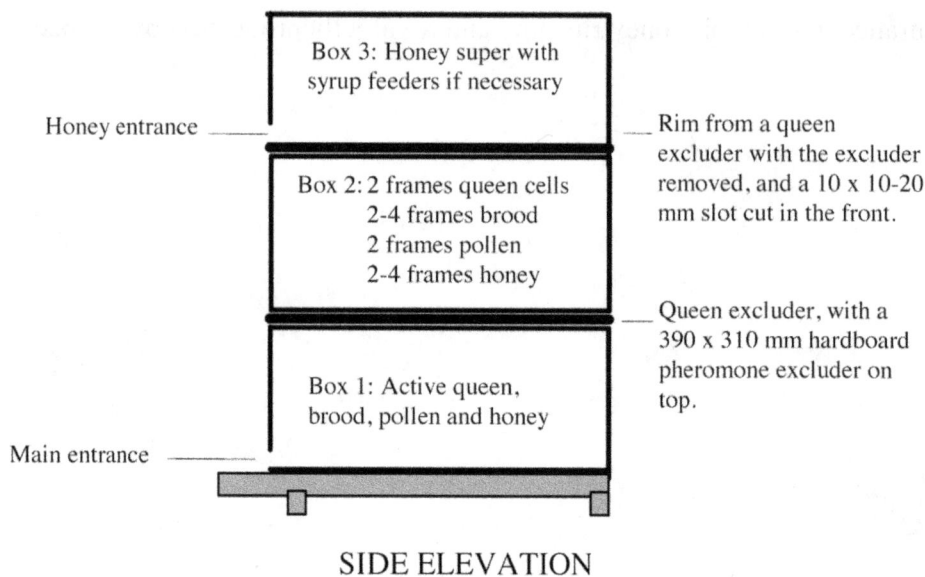

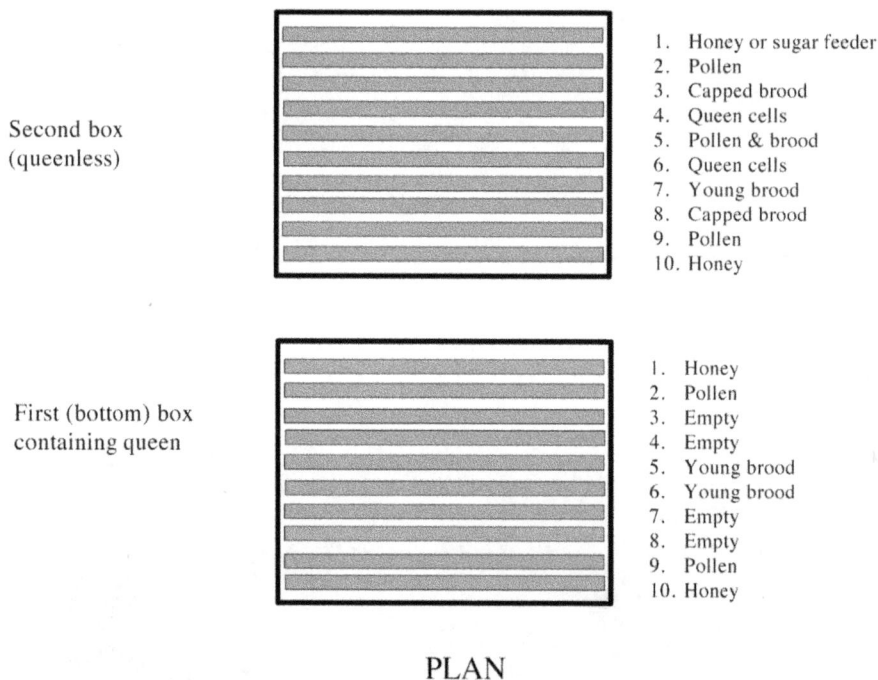

Figure 3: Arrangement of a hive for royal jelly production.

4.6 Queen cell frame

The plastic queen cells are arranged in the middle of a three-quarter or a full-depth frame in two groups of 15 cells. The top group is positioned immediately below the top bar and the lower group positioned so that the entrances of the lower cells are 100 mm below the entrances of the top cells (Figure 4). Two frames containing 30 cells each should be grafted for each hive. In full-depth boxes, three bars containing 15 cells each can also be used.

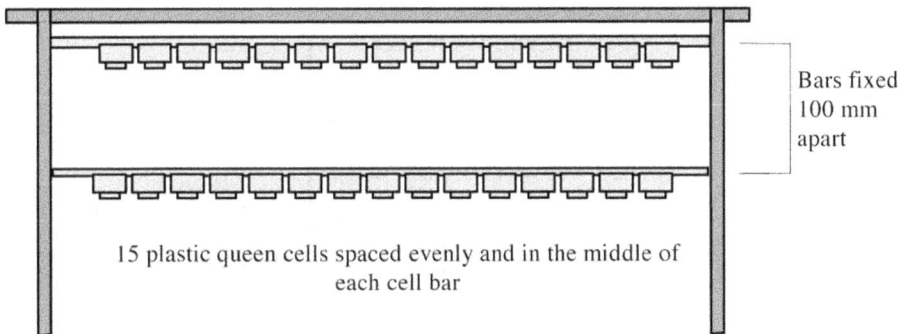

Bars fixed
100 mm
apart

15 plastic queen cells spaced evenly and in the middle of
each cell bar

Figure 4: 30 queen cells arranged on the cell bars for three-quarter or full-depth boxes.

4.7 Nucleus hives

The hives that are used to provide newly hatched larvae for transfer to queen cells, should be specifically set up as single story hives, which are referred to as nucleus hives (nucs) in this text. It is quicker and easier to take the larvae from these nucs rather than from the surrogate hives or from frames in the bottom box of hives being used for royal jelly production. These nucs should be configured to give easy access to the young larvae (Figure 5). They also require a one-year old queen because of the demand for newly hatched eggs when royal jelly production is under way. An older queen will be less able to satisfy the high egg laying requirement. If a queen lay cage (Page 49) is to be used, an empty frame will need to be removed to accommodate the extra thickness of the cage.

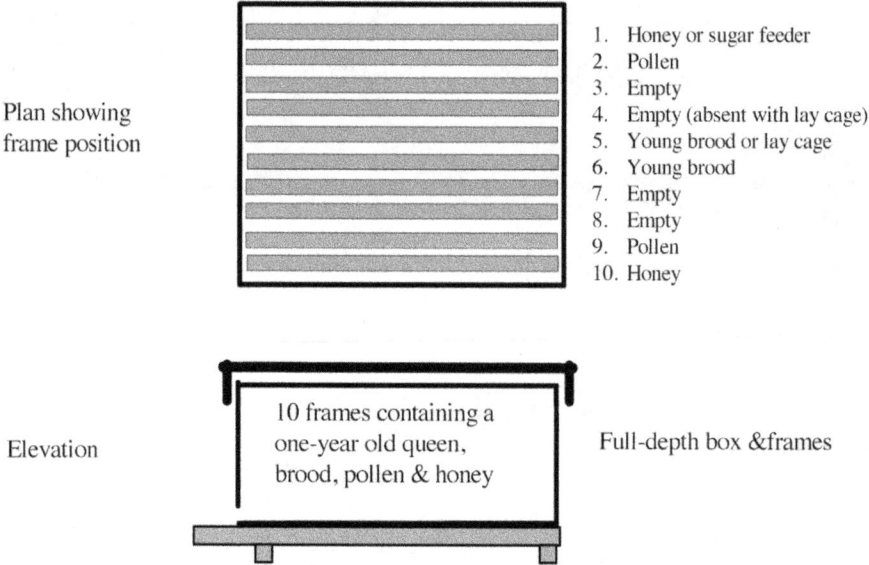

Plan showing frame position

1. Honey or sugar feeder
2. Pollen
3. Empty
4. Empty (absent with lay cage)
5. Young brood or lay cage
6. Young brood
7. Empty
8. Empty
9. Pollen
10. Honey

Elevation

10 frames containing a one-year old queen, brood, pollen & honey

Full-depth box &frames

Figure 5: Arrangement of nucs for the production of newly hatched larvae.

4.8 Colony strength

The strength of the colony largely determines the quantity of royal jelly produced from each hive. Nutrition, health and genetic characteristics of the colony influence jelly yields to a lesser extent. Colonies should have at least 25,000 bees at every three-day grafting cycle.

To estimate the number of bees present, determine the amount of bee cover over each frame either in the morning or evening when most of the bees are in the hive. A full-depth frame with no comb exposed contains about 3,500 bees (Plate 5). A similarly covered three-quarter depth frame contains about 2,300 bees. A half-covered frame will have half these numbers of bees on it. Simply add the estimates of bees on each frame to obtain the number of bees in a hive.

Bolster weak colonies with worker bees from surrogate hives. Either shake a frame with adhering bees over the weak colony, or by replenishing frames in the second box with frames containing bees, brood and pollen.

Ten-frame boxes containing bees, brood and pollen from surrogate hives can be 'papered' onto very weak colonies to create strong ones. This method involves condensing all the bees and brood from the weak hive into one box immediately above its bottom board. To do this, replace all empty frames in the bottom box with brood, pollen and honey frames from the other boxes of the weak hive. The frames should be arranged as for the first box of the hive configuration in Figure 3.

Place two layers of newspaper over the box. Add a box from a surrogate hive complete with bees, brood, pollen and honey as the second box. Take particular care in ensuring that the queen is absent from this box. If she is present, place her in the bottom box of the surrogate hive. Add the honey super onto the 'new' hive as in Figure 3. The bees from both the first and second boxes will chew through the paper within 48 hours and merge together as one harmonious colony.

4.9 Finding queens

4.9.1 Searching method

In royal jelly production, keeping queens out of the box containing the plastic queen cells used for jelly collection is crucial. Production will cease immediately if a mated or virgin queen is present. Pheromones (volatile compounds which affect the behaviour of other bees) emitted by the queen at a high enough concentration will inhibit the production of new queens. Workers which detect queen pheromones, either by physical contact with the queen, in the air or on the wax comb, will not feed royal jelly to larvae in the plastic queen cells. Instead, they will tear the cells down, killing the larvae. Therefore, it is important to be able to detect queens easily when you are working with the hives.

When looking for the queen, move slowly but positively, taking care not to jar the hive. Use a minimum amount of smoke so as not to stampede the bees. A mated queen (Plate 1) can be recognised from worker bees by her long abdomen and slow deliberate walk. The queen will often have a court of workers surrounding her in a circle. Virgins have smaller abdomens than mated queens, and scurry rapidly around the frames, making them much harder to find. When inspecting the frames, scan the comb face closest to you, inspecting the edges in case the queen is moving around to the other side. Then scan the middle of the face. Flip the bottom of the frame away from you so the frame is upside down, and the second side is towards you. Check the edges of the frame as it is flipped over. Inspect the frames over the box so if the queen accidentally drops, she will be contained safely in the box. The queen is most likely to be found on the middle frames of the second box of a hive where the brood is spread over the two bottom boxes.

To find the queen in a double-box brood chamber of a full-depth or three quarter-depth hive prior to setting up the hives for royal jelly production, put the second box on an upturned hive lid. Cover the bottom box with the inner cover to reduce the risk of bees from other hives robbing the honey. Standing along side the second box, remove the outside frame closest to you, inspect it for the queen, and place it near the hive entrance. If an undetected queen is on the frame, she can walk back into the hive to be detected later. Repeat for the next frame in, sliding the frame towards you before lifting it up to avoid crushing any queen present between the vertically moving and stationary frames. Place the second frame against the first frame at the entrance of the hive. A gap of the width of two frames will now exist in the box. Before looking at the second frame, glance at the face of the next comb and to the sides and bottom of the box for the queen. Slide the third frame in towards you and lift it out, inspect it for the queen, and place it hard against the side of the box. Once the fourth frame has been inspected, place it hard against the third frame. Repeat this procedure for the other frames until all the frames have been thoroughly inspected. Place the last frame back in the original position and using your hive tool as a lever against the side of the box closest to you, slide all the frames together away from you. This can be done with Hoffman end bars but frames with Simplicity end bars need to be spaced individually. Place the first two outside frames back in their original position.

If newly laid eggs are present in the second box (excluding brood in frames bought in from other hives), the queen will most likely be present there somewhere. If the queen is found, place her in the first box either by shaking the frame over the box, or by picking her off the frame by her wings with your thumb and forefinger. If the queen has not been found, you can tentatively assume that she is in the bottom box or that you missed her. Either way, assemble the hive as described on Page 25. Search in the second box for the queen again to ensure she is not there before royal jelly production commences.

If the brood chamber is made up of three three-quarter depth boxes, put the queen in the bottom box below a queen excluder, with the top two boxes queen-less for royal jelly production.

4.9.2 Marking queens

Finding a queen can be made easy by marking her on her thorax with a drop of aero-modeller's enamel, automotive lacquer (e.g. car touch-up paint), typist's correction fluid or nail polish. Apply the paint either with a fine brush or matchstick as she walks across the comb or while you gently hold her still on the comb between thumb and forefinger.

For consistency, I recommend using the international colour code for recording the queen's age (Table 4). Each year is assigned a different colour on a five-year cycle.

Table 4: International colour codes for marking age of queen.

Year ending with:		
	1 or 6	White
	2 or 7	Yellow
	3 or 8	Red
	4 or 9	Green
	5 or 0	Blue

4.10 Supplementing the bee diet

Prior to, or during royal jelly production, the colonies may require protein or sugar substitutes or supplements in addition to what they can gather naturally. With prudent timing these food additives can build up bee numbers to improve production without reducing the natural quality of the royal jelly.

4.10.1 Protein feeding

Under royal jelly production, colonies may run short of pollen especially in spring when demand for protein is particularly high and bad weather restricts foraging. Pollen deficiency is likely if pollen reserves in the hive dwindle to less than half a comb. For hives harvested for more than two weeks, protein supplements or substitutes are recommended.

4.10.1.1 Protein substitutes

Pollen proteins can be substituted with dairy products, which can be obtained from a milk processing factory, and deactivated brewers yeast.

The best recipe currently available is the 'Beltsville Bee Diet' developed by the United States Department of Agriculture.

A 1 kg patty consists of (on a dry basis):

1. Lactalbumin or alternatively caseinates   120 g

2. Deactivated brewers yeast   230 g

3. White sugar   650 g

4. Water to half the weight of the sugar, i.e.   325 ml.

Divide the dough into 500 g balls and roll to a thickness of 10 mm. Store patties in a domestic freezer at −15°C between 200 x 200 mm sheets of wax paper.

A 500 g protein patty fed every six days can increase royal jelly yields in pollen-deficient areas by an average 2.5 g/hive/harvest (36%) over a month of production.

This protein substitute will not influence the quality of royal jelly as indicated by the standard quality factors of colour, moisture, protein, carbohydrates, lipids, ash and 10-hydroxy-2-decanoic acid. However,

Lactalbumin and deactivated brewers yeast may be hard to obtain. Ready-mixed proprietary protein substitutes are available commercially and may be a better option.

For simplicity, patties can be put on top of the queen-cell frames when hives are arranged for royal jelly production, and every 6 days thereafter to coincide with harvest times. The patties will be fully consumed after this time. However, the position of the patties can be important. Compared with placing the patties on top of the frames, hanging them between brood frames has been recorded to give 20-35% improvement in royal jelly yield.

This yield increase is due in part to an increase in cell acceptance and an increase in hypopharyngeal gland size of newly emerged worker bees. These young workers become nurse bees within six hours of emergence and feed royal jelly to the developing queen larvae. They remain in the warm brood area where they emerge. The nurse bees clean empty cells for queen egg laying and feed on pollen processed and stored by older bees just near the brood site. Consequently, they have easy access to an alternative diet if pollen supplements are hung between the brood frames. Much of the ingested pollen is converted to worker jelly or royal jelly. The bees gradually move out of the brood area only when they are older. Protein supplements placed on the bottom board or top frames tends to be consumed by foraging bees rather than the young nurse bees. Moreover, direct transfer of food to older worker larvae is less frequent in top-frame feeding than in brood-area feeding, with the shortage of food supplemented by worker jelly. Therefore, brood area feeding provides a more efficient nutrient flow to young and old queen larvae.

Brood-area pollen supplement feeders are being developed for commercial use, but currently their availability is limited. Smearing a wet protein supplement mix with a flat spatula over empty honey frames is a practical and cheap alternative, but the effectiveness of this feeding method in improving royal jelly yields has not been measured by the author. Put only a small amount of the protein supplement in each cell, so that when the frame is inclined vertically in the hive, the mix will not flow over the lip of the cell. Smear the supplement in the middle 150 mm of the frame only. To prevent the brood area being split in two, cut out the wax in the frame either side of the supplement so bees have ready access around the feeder to the queen cells. Place one frame containing the supplement near a queen cell frame.

4.10.1.2 Pollen supplements

Feed pollen mixed with two parts sugar syrup to one part water in a side feeder placed against the inside of the second box. Use 200 g pollen for 10 litres of sugar syrup. The pollen can be collected using a pollen trap fitted to surrogate hives in times when pollen is in abundance, e.g. in summer. The sugar syrup is made by dissolving 6.6 kg sugar in 3.4 litres of hot water. The patties need to be stored in a freezer for at least one month to kill wax moth eggs and larvae. Wax moths damage comb, although they are usually controlled to low levels in strong bee colonies.

Feeding pollen has two possible disadvantages:

1. Diseases such as American Foul Brood (Page 79) are easily spread in the pollen to the royal jelly producing hives.

2. The sugar syrup, being liquid may be fed directly to the queen larvae, and as a consequence alter the chemical constituents of royal jelly.

Thus, this method is not favoured unless you are certain that no disease is present in any hive.

4.10.2 Sugar feeding

In addition to feeding of protein patties, sugar feeding is necessary in early spring if the honey supply in the colony is below that contained in four, full three-quarter depth frames. The colony will not usually be able to build up satisfactory bee numbers with any less honey. Sugar syrup stimulates the colony to expand in bee numbers. Feeding sugar at least three weeks prior to commencing royal jelly production will ensure that most sugar will be used in brood rearing and not contaminate the royal jelly. Feed only honey as the source of sugar to colonies three weeks before and during royal jelly production to ensure the royal jelly contains only honey-derived sugars.

Sugar syrup can be made by dissolving white sugar or raw sugar in an equal weight of water. Avoid using brown sugar or molasses as these cause dysentery in adults. The syrup can be made by filling a container three quarters full with sugar, topping it up with boiling water, and stirring at intervals until all the sugar dissolves. Feed 6 litres to each hive.

Feeding sugar syrup in autumn to sugar-deficient colonies after royal jelly production has been concluded for the season is recommended. The sugar will be converted by the bees to stores to over-winter on. In making the syrup, increase the concentration of the sugar from that used for spring feeding to two parts sugar to one part water (by weight). The colony will store in the comb a little less than the weight of dry sugar in the original syrup. Weaker syrup concentrations tend to be consumed directly as the colony is stimulated to increase brood. As above, feed 6 litres to each hive. Syrups can be fed in a frame feeder replacing an outside honey frame in the second box. The frame feeder can be made from plastic, waxed-sealed plywood or tin nailed to a wooden frame. Place floats such as bracken, fern, wood-shavings or gauze in the feeder to prevent bees drowning in the syrup. Another practical feeder system is the top feeder. It comprises a half-depth box with a hardboard bottom fitted into a saw cut 8 mm up from the bottom to provide a bee space over the tops of the frames in the box below. Bees have access to the syrup from the hive below via a 'chimney', a solid block of wood with holes bored in the middle. As for the frame feeder, flotation material is required for this system.

4.10.3 Water

Water is an absolute necessity in the bees' diet. It is used mainly in brood rearing by diluting larval food and for cooling when temperatures in the hive exceed 35°C. Bees collect water from moist soil or droplets on leaves. If no water is available for the bees to collect in summer when demand for water is high, drip water from a partially opened tap over a mound of sandy soil or bark chips for the bees to walk on.

4.11 Time of year

The ideal time for production of royal jelly is in late spring. Honeybee colonies are large in preparation for the honey flow. They are preparing for queen rearing and supersedure. Pollen is abundant. Royal jelly production can continue for 6-8 weeks before the honey flow, but may have to cease then because bees have a tendency to deposit honey into the queen cells. Royal jelly production can resume again in late summer-early autumn, immediately after the honey flow, which is also a good time for jelly production. Production can last 3-4 weeks before jelly yields reduce gradually to low levels as natural pollen and honey sources become rare and the colonies prepare for the winter.

4.12 Summary of equipment

Below is a list of equipment required for the production and packaging of fresh royal jelly. Most of this equipment can be purchased from bee equipment stockists.

1. For every six royal jelly producing bee hives you need three surrogate hives and two nucs.

2. A vehicle. This can be used as shelter for grafting if alternative shelter is unavailable.

3. A laboratory e.g. kitchen, containing a stainless steel sink and bench, hot water for washing equipment, electricity and good

natural light for harvesting, weighing and storing of jelly and bee larvae.

4. Weighing balance accurate to $\pm$ 0.05 g, used for weighing fresh royal jelly into glass vials and recording yields accurately.

5. Suction apparatus and syringe-sieve unit (Figure 6) for harvesting royal jelly.

6. Vacuum pump. The battery powered type used for inflating rubber rafts is suitable. Another alternative is to use the air blowing outlet on an electric vacuum cleaner.

7. Freezer capable of storing royal jelly to -15°C for periods up to three months, or liquid nitrogen to freeze the jelly at -196°C to prevent any chemical deterioration for longer-term storage of royal jelly.

8. Suitable glass vials for storage and sales of royal jelly e.g. 50 mm x 19 mm external diameter flat-bottom glass tubes with a poly-stop cap hold 10 g of fresh royal jelly.

9.   Protective clothing: cuffed overalls with attached hood and veil, gloves and gumboots.

10. Head lamp, smoker and hive tool.

11. Larval transfer aids:

     i.   Retractable spring-loaded grafting tool, or,
     ii. Artist paint brush size '00' or '000', or,
     iii. Queen lay cage larval transfer systems.

12. Plastic queen cells and full-depth frames (two per hive), 30 cells per frame (Figure 4).

13. Hive marking pens and forms for recording harvesting details.

14. Pollen supplements: lactalbumin, deactivated brewers yeast and sugar in a ratio of 1:2:5.4, respectively (Page 36).

15. Empty full-depth 'holding' box with a fixed bottom and hinged lid. This box is used to convey frames of queen cells filled with royal jelly from the production hive to indoors for harvesting.

Plunger to force harvested jelly
through sieve once pooter is removed

To suction pump

2 mm
internal

4mm
internal

Rubber or plastic
bung

80 mm

30mm internal diameter
(approx.) 60 ml plastic

Sieve

Stainless steel sieve (24,
0.18mm wide strands/cm)
with aluminium flange

Air-tight cap

Diameter to
neatly fit syringe

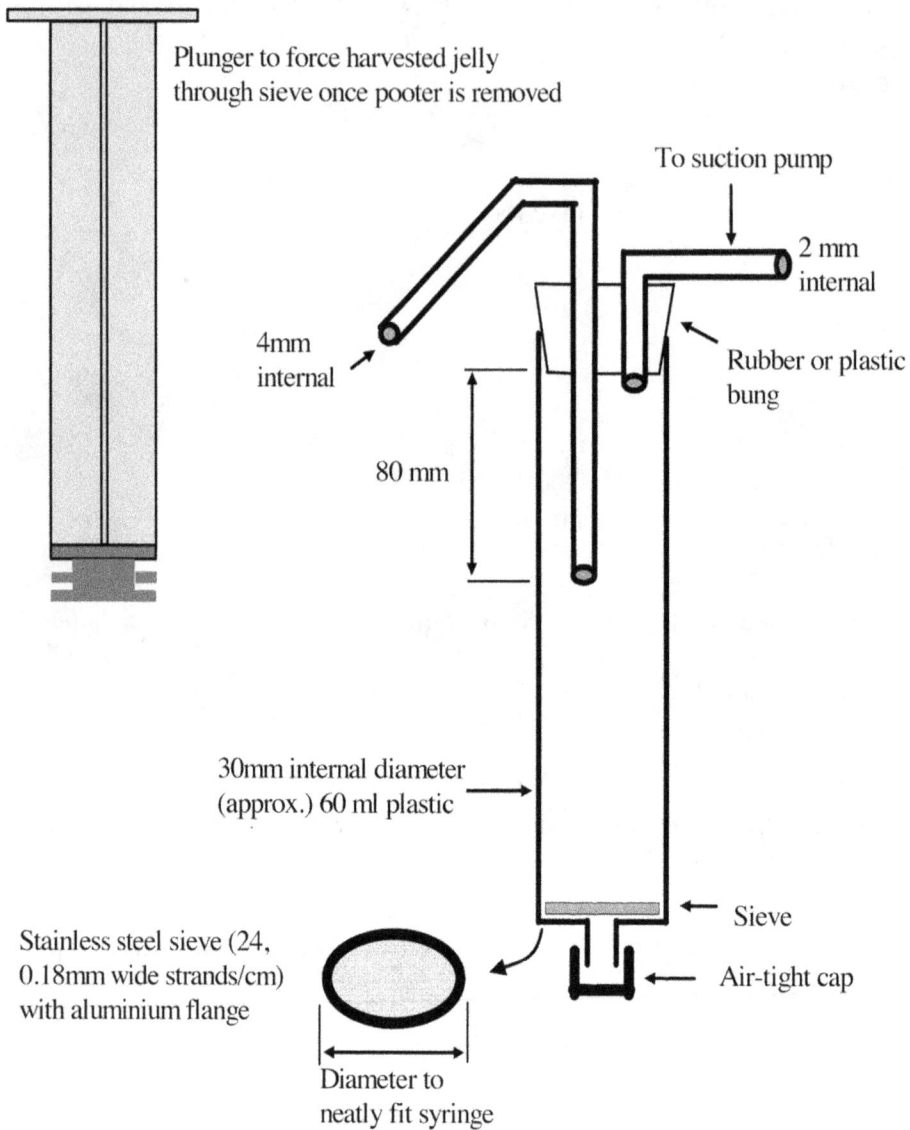

Figure 6: Suction apparatus for harvesting royal jelly.

5.   Larvae Transfer

There are two basic techniques available for transferring newly hatched brood from where the queen laid them to the queen cell cups, each with advantages and disadvantages.

5.1 Grafting

In this method, newly hatched eggs (Plate 6) are transferred to the queen cells using either a retractable spring-loaded grafting quill (Plate 7) or a fine artist's brush, size 00 or 000. The preference for using either the brush or tool is personal, although the retractable grafting quill will probably cause less damage to the delicate larvae for beginners. It is also more likely to collect most of the jelly surrounding the larvae, an important factor in ensuring its survival.

Remove the frame containing the young brood from the nucs or hives supplying the brood, shaking the bees off the frame onto the colony. Place the frame that is mostly free of bees on a bench under adequate lighting, but not direct sunlight as this may kill the larvae. A battery operated lantern or torch attached to your bee hat is ideal. The bee veil may have to be removed to enable you to see the small 2-6 hour larvae. With the top bar towards you, incline the frame containing the young larvae upwards at about 20° so you can look directly down, with illumination from the hat-mounted lantern, into the bottom of the cells (Plate 6).

To remove a larva from a worker cell, slide the tool or brush down the side of the cell closest to you, so the tip just curls under the larva. Lift the implement straight out of the cell, bringing with it the larva and as much worker jelly as you can. As the grafting implement is lifted upwards, press it against the cell to leave an indent in the wax. This mark indicates where the larvae have been removed, thereby avoiding wasting time in attempting to remove larvae from empty cells. Place the larva into a queen cell cup, and either roll the brush slightly or retract the tip of the tool to pull it out from underneath the larva. Discard any larva that has rolled as the adhering jelly will probably have blocked its breathing holes (spiracles). Also discard any larva that has been brushed against the cell wall, as it will almost certainly not develop.

The grafting method has the advantage over the lay cage systems in that little preparation is required to select young larvae. No outlay in extra equipment is required, and for experienced operators, grafting can be very quick. However, the grafting tools may damage the delicate larvae in the transfer to the queen cells. If that occurs, the larvae die and worker bees will not deposit royal jelly in those cells.

5.2 Queen lay cage

Queen cage systems sold under brand names 'Nicot', 'Jenter' and 'Ezi Queen & Royal Jelly "Larvae" Transfer System' are available. These systems involve placing a queen into a cage positioned in the centre of a nuc (Figure 5) or in the brood chamber of a beehive (Figure 3). The queen lays eggs within 24 hours onto plastic dimples. These are removed by hand after three days and transferred to the pre-prepared queen cells.

The cage systems involve the one-off preparation of fitting the cage to a frame, lightly waxing the front dimples and smearing the dimples with sugar syrup to attract bees to it. The cage is placed into a specially prepared nuc for the bees to draw out the comb (Plate 8). The comb is drawn out usually in 1-2 days. The queen is then caught by placing the thumb and forefinger around her wings. Once caught, she is placed into the cage (Plate 8 inset). After three days, the plugs containing the newly-hatched larvae are transferred to the queen cell bars (Plate 9). The queen cells will be drawn out with wax if the larvae they contain are accepted by the bees (Plate 10).

The difference between the brands of cage systems lies in the ease and speed of transferring the newly hatched larvae from the lay cage to the queen cells, and in the number of eggs that can be laid by the queen in the cage at any one time. The 'EZI Queen system' incorporates ingenious bi-plug plastic strips, which dove-tail together when fitted to the rear of the lay cage to give a compact laying pattern for the queen. The compactness of the design allows for 420 eggs to be laid at a time by the queen. Unlike the other systems where only the individual eggs can be transferred, these bi-plug strips in the 'EZI Queen system' are later split in two. This arrangement enables the newly-hatched eggs to be transferred to the queen cell cups, 10 at a time (Plate 9 inset), equivalent to about 90 cells per minute. The queen cells are pre-aligned on the cell bars to fit the plugs.

Cage systems have the advantages over manual grafting of being able to transfer 2-6 hour old larvae with reduced chance of injury. They do not rely on perfect eyesight and good hand-eye co-ordination for their success, and they allow for the use of gloves to reduce the chance of being stung.

Disadvantages include the comparatively high capital cost, and the initial set up time (1-2 hours per cage) involved in mounting the lay cage in a comb frame to allow worker bees to draw out the cage cell templates with wax. Other jobs included in the set-up time involve locating the queen so she can be placed in the cage, and releasing the queen after she has laid an

egg on most of the dimples in the cage.

Plate 6: Selection of 2-6 hour old larvae (inset) suitable for grafting, inclining the frame so the cells point directly upwards.

Plate 7: Retractable grafting quill used to transfer newly hatched larvae to the queen cells.

Plate 8: Workers drawing out comb in a lay cage. A queen being put into the cage (inset).

Plate 9: Transferring the plugs containing young larvae (inset) from the lay cage to the cell bars using the 'EZI Queen larvae transfer system'.

6.   Harvesting

Prior to harvesting, remove the honey super and lift the queen cell frames out of the second box. Holding the frame at both ends, shake off the adhering bees over the hive by giving the frame a sharp downward jerk. This jerking method of removing the bees is preferable to the slower and more damaging alternative of brushing bees off the frame using a soft-hair paint or proprietary bee brush. The cells drawn out with wax will contain royal jelly (Plate 10).

Place the frames in a sealed container (holding box). Ensure that the frame is labelled so as to correctly identify the frame as coming from that hive. This limits the risk of transmitting diseases between hives in the event of frames being mismatched to their original hives. It is more efficient to remove all the queen cell frames from the hives before commencing harvesting, rather than harvesting each queen cell frame individually immediately after removal. All the frames can be harvested in succession away from the hives and without the interference of bees.

In the 'laboratory', hold the frame upside down. Either open out the entrance of the waxed cell to expose the larva with tweezers, or remove the wax around the cell entrance with your fingers. Remove larvae from the cells using tweezers (Plate 11). The wax can be collected and purified in bulk later. The larvae can either be discarded into a plastic-lined bin which is sealed to prevent disease transfer, or immediately frozen for sale. There is a market in Asia for queen larvae.

Harvest the jelly from the cells by vacuum using a suction apparatus (Plate 12). The amount of jelly that can be harvested from 44 queen cells is shown in Plate 13. With the 60 cells recommended, greater yields can be expected. Once all the jelly is harvested from the frame, place the frame in the holding box ready for re-grafting. Immediately strain the jelly in the syringe into glass or plastic vials, or in bulk containers, and freeze it. The rubber on the syringe plunger will need to be protected from the denaturing effects of royal jelly. To achieve this, the rubber can be sprayed regularly with a food-grade silicone sealant.

6.1 Colony selection for yield

Of the hives used for production, there will be colonies that produce yields of royal jelly higher than the average. High-yielding colonies include those with large colony size, a high proportion of nurse bees, dwindling food stores, and a favourable genetic disposition towards producing royal jelly. To maintain or improve royal jelly yields from an apiary, I recommend that after the third harvest cycle the worst 10% performing colonies be removed altogether from production, and replaced with hives from the surrogate group. These replacement hives will need to have been prepared as described (Page 25) to accept newly-hatched larvae at the same time as those of the main production group.

6.2 Harvesting royal jelly and honey

If bee colonies are harvested for royal jelly in spring to early summer, and converted to honey producing hives immediately prior to the onset of the honey flow, surplus honey may be collected by the bees for harvest. However, if royal jelly production continues through even a portion of the main honey flow, little surplus honey can be expected. All the nectar collected by the bees will be used by the bee colony for its own use and for producing royal jelly. Consequently, the value of honey lost must be budgeted for when calculating the financial returns from royal jelly.

A colony used for royal jelly production can be easily converted to a honey producer. After a harvest of royal jelly, simply:

1. Remove the pheromone excluder between the bottom and second box.

2. Place frames of empty laying comb in the second box where the queen cells used to be.

3. Place a queen excluder over the second box.

4. Place one or more honey supers on the queen excluder.

5. Place a hive mat and lid on the top box.

The hives can be converted back to royal jelly producers after the full honey supers have been removed for harvesting at the end of the main honey flow. Simply:

1. Put the queen, which is most likely to be in the second box, in the bottom box.

2. Place the queen excluder on the bottom box, and place the pheromone excluder on top of the excluder bars.

3. Replace empty frames in the second box with one or two queen cell frames containing young larvae.

4. Harvest the royal jelly in 72 hours.

6.3 Alternative cycle periods

The optimum duration between grafting or larval transfer and harvest appears to be flexible. The author found no significant difference in royal jelly yields occurring between periods ranging from 66, 72, or 78 hours. For convenience, aim to harvest 72 hours after larval transfer, knowing that you can harvest six hours either side of the time of larval transfer without suffering any yield loss. That is, you can harvest at any time on the third day without affecting royal jelly yields.

Durations outside the range of 66 to 78 hours after larval transfer appear to result in lower royal jelly yields in relation to the effort required. Harvesting royal jelly every 48 hours has been reported to give greater yields over a month than harvesting every 72 hours after larval transfer. Unfortunately, the extra work in grafting or larval transfer every two days instead of every three days tends to negate the returns obtained from extra yield.

A technique of harvesting royal jelly twice after every larval transfer to save time involved in grafting or using the larval transfer systems is not recommended. This technique is referred to as the double harvest per larval transfer. With this technique, royal jelly is harvested 72 hours after larval transfer as with the single harvest per transfer regime, but instead of discarding the larvae, the larvae are carefully replaced in the queen cells immediately after harvest. The frame is placed back in the hive to allow the bees to deposit more royal jelly into the cells. The newly deposited royal jelly is harvested again 48 hours later after the five-day old larvae have been removed and discarded. Newly hatched larvae are then grafted into the empty cells to complete the cycle. In trials however, the extra time taken to carefully remove the larvae without damaging them prior to harvest, and placing them back in the cell after harvest, negated any time saved in grafting or larval transfer. Any damaged larvae meant no jelly would be deposited into those queen cells. Further, yields were half that obtained from a single harvest per graft or larval transfer for an accumulated production over one month. The poor yields were attributed to the 3-5 day old larva consuming extra royal jelly fed to them by workers prior to the second harvest, compared to the smaller 1-3 old larva used in the first harvest.

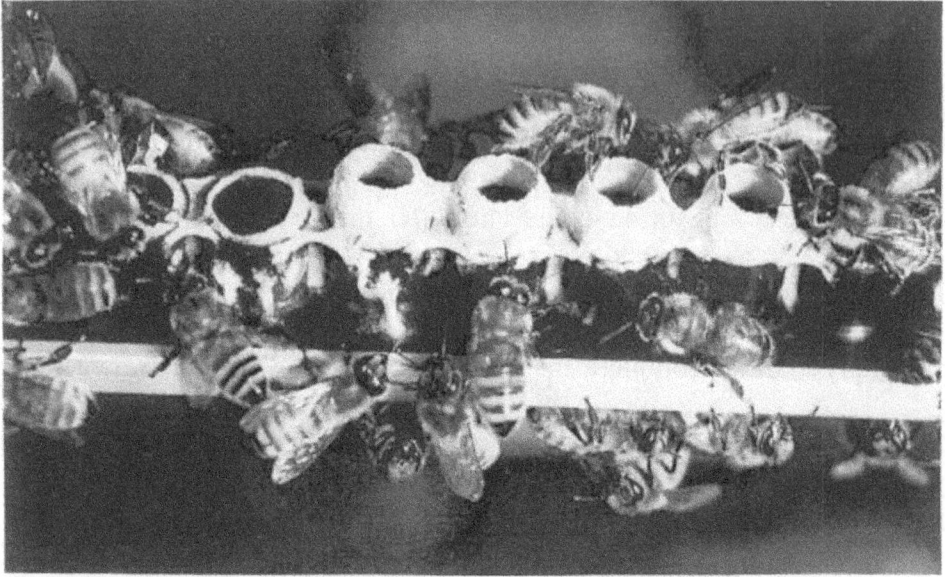

Plate 10: Wax drawn out at the entrance of accept cells.

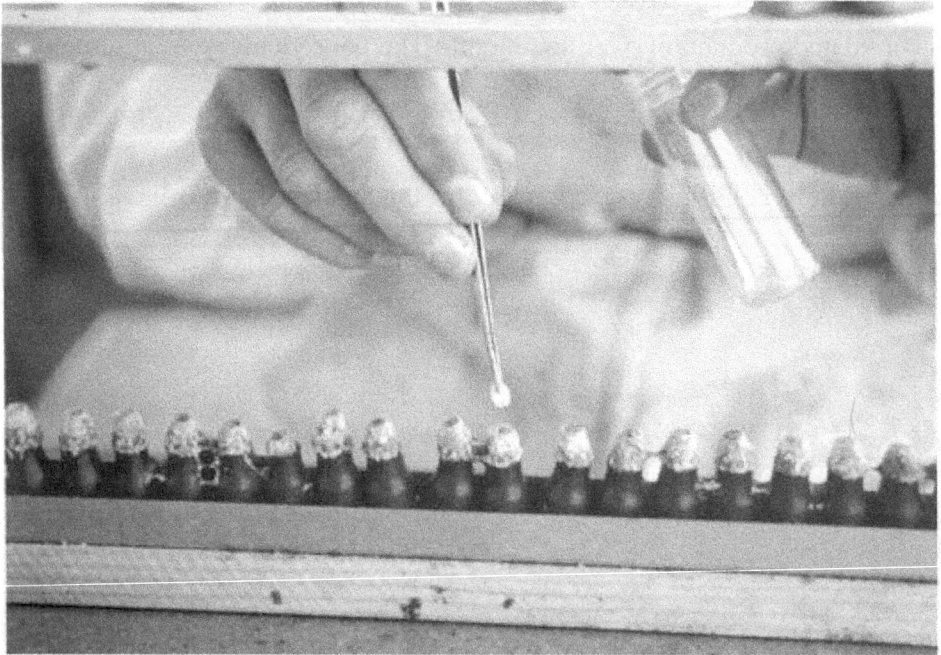

Plate 11: Discarding larvae in preparation for harvesting.

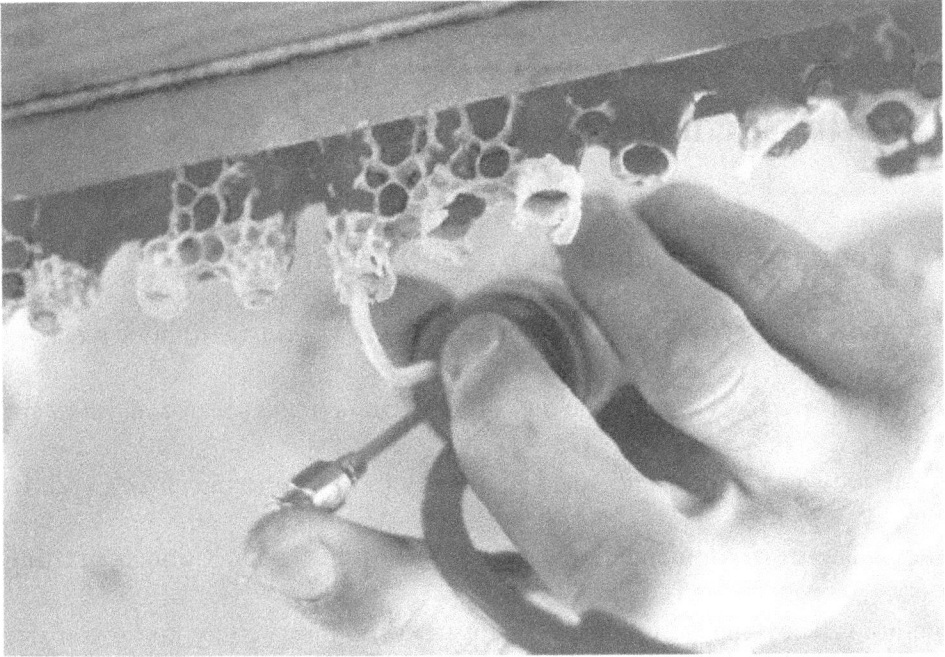

Plate 12: Harvesting royal jelly with a suction apparatus.

Plate 13: 10 g royal jelly harvested from 44 queen cells.

7.   Management Procedure

7.1 Commercial production by grafting larvae

The steps itemised below allow one beekeeper, working a 40 hour week on a cycle of two days on and one day off, to maintain and service 30 hives for royal jelly production and the accompanying 15 surrogate hives and 10 nucs. These recommendations apply to using the traditional grafting methods. They can be modified easily to accommodate any larval transfer system.

7.1.1 Establishment

THREE WEEKS prior to royal jelly production check the hives intended for this purpose for honey stores. Feed sugar syrup to every hive which has less than four full frames of honey (Page 40).

FOUR DAYS before grafting, arrange 30 strong hives containing more than 25,000 bees in a semicircle, with entrances facing outwards (Page 23, Figure 2). The 15 surrogate hives and 10 nucs should be sited more than 100 meters away from the main apiary to limit transferred bees drifting back.

Manipulate the ten 10-frame nucs to include a recently mated or one-year old laying queen, two pollen-honey frames, four empty frames in the middle of the box for the queen to lay in, two brood frames and two honey frames (Page 30, Figure 5). Feed the nucs with 100 g of protein patties placed on top of the frames.

TWO DAYS before grafting set up 15 hives for royal jelly production as described on Page 25. Place a one-year old queen along with four empty frames, two frames of pollen, two frames of honey and two frames of brood in the bottom box. Add the queen pheromone excluder above that. Add the second box containing a pollen frame sandwiched between the two frames of queen cells in the middle of the box. On either side of the queen cells insert one frame of capped and uncapped brood, one frame of pollen and one frame of honey. Additional brood frames can be included to make up 10 frames, particularly if only one queen cell frame is used. In periods of a honey flow, add the alternative honey entrance above the second box and place a honey super on top. A hardboard hive mat and water-proof lid is necessary for insulation. Repeat this procedure for the remaining 15 hives the following day.

7.1.2 Production maintenance

Day 1.   If feeding supplements, in the morning place 500 g protein patties onto the middle of the pheromone excluder of each hive intended to receive the grafted larvae.

Starting at about 1 pm, graft 6 hour-old hatched larvae from the nucs into 15 prepared hives (Batch A).

Day 2.   At 1 pm graft 2-6 hour old larvae from the nucs into the other 15 (Batch B) prepared hives.

Day 3.   Rest day.

Day 4.   At 8 am remove queen cell frames from Batch A and place them in the holding box (Page 54). In the laboratory, remove larvae from cells by tweezers (Page 60, Plate 11) and discard, or save for human consumption.

Harvest jelly from the cells by vacuum using the suction apparatus (Page 46). Strain the jelly in the syringe, pack in glass vials or in bulk, and freeze immediately.

At 1 pm graft 2-6 hour old larvae from the nucs into the 15 Batch A hives.

Day 5.　　At 8 am remove queen cell frames from Batch B. Remove larvae and harvest. Pack the royal jelly in glass vials or in bulk and freeze it immediately.

Day 6.　　Rest day.

Day 7 and beyond.

Repeat from Day 4, harvesting and grafting Batch A hives. Inspect nucs and ensure a clean empty frame is available for the queen to lay in.

If feeding protein supplements, in the morning of every second harvest cycle, add a 500 g protein patty to each hive and 50 g to each nuc.

In the morning of every third harvest cycle, boost royal jelly producing hives with frames of brood, adult bees, pollen and honey as necessary from the surrogate hives. Ensure that foreign queens are not accidentally introduced into the royal jelly producing hives.

Check the frames in the second box for rogue queen cells every six days. Cull any queen cells present.

Replace filled honey supers as required with empty honey supers.

At every harvest check the hives for royal jelly yield and cell acceptance. Hives producing yields below 180 mg/cell and with a cell acceptance below 60% should be replaced with the surrogate hives which have been previously configured for royal jelly production. The surrogate hives need to be at a stage where they can accept newly-hatched larvae in the queen cells.

After the third harvest, replace 10% of the lowest performing hives with surrogate hives to maintain and possibly improve yields from the apiary.

7.2 Commercial production using larval transfer systems

As before, the steps itemised below allow one beekeeper working a 40 hour week to maintain and service 30 hives for royal jelly production and the accompanying 15 surrogate hives. But instead of 10 nucs as in the grafting method, only five nucs are required for the queen lay cages since the lay cages will allow the queen to lay enough eggs to service a ratio of one nucleus hive to six royal jelly hives containing 60 queen cells each. The three days it takes for the eggs to hatch after the queen has laid them in the dimples coincides with the three-day cycles in royal jelly production.

Because setting up the queen lay cages in the nucs is so time consuming in relation to larval transfer, I recommend that for all 30 hives, royal jelly harvest and the transfer of young larvae to harvested cells is conducted in one day, rather than to roster the harvest over two days with two groups of hives as recommended with grafting. This reduces the preparation time required to manipulate the nucs. The following instructions apply particularly to the split plug type lay cage system which allows for a large number of eggs to be laid in each cage set (Page 53). Nevertheless, the instructions can be easily adapted to cater for the other queen lay cage systems commercially available. These instructions enable a beekeeper to work the 30 hives on a roster of one day on followed by one and a half days off.

7.2.1 Establishment

THREE WEEKS prior to royal jelly production check the hives intended for this purpose for honey stores. Feed sugar syrup to every hive with less than four full frames of honey (Page 40).

Snap the plugs into five lay cages, attach the rear cover, and coat the lay cage with a thin layer of warm bees wax. Insert the assembled cage into a frame so that the cage is held in the frame, surrounded by drawn wax or foundation (Plate 8). Replace an empty frame in a surrogate hive with a frame containing the uncovered queen lay cage to allow the workers to draw out the lay cage comb. Arrange the frames so that the frame containing the lay cage is in the middle of the second box.

THREE DAYS before egg transfer, arrange 30 of the strongest hives in a semicircle (Page 23, Figure 2). Locate the surrogate hives and nucs more than 100 meters away from the main apiary to limit transferred bees drifting back.

Manipulate the five nucs for production of young larvae to prime the queen cells for royal jelly production. Transfer the frame containing the queen lay cage from a surrogate hive into the middle of the nucs. Fix the cover onto the lay cage and put the queen inside (Plate 8). Arrange the nuc frames so that the frame containing the lay cage is adjacent to one brood frame in the middle of the box. Either side of these two frames, working to the outside position of the box, place two empty frames, one frame of pollen/honey, and an partial frame of honey on the outside for honey consumption by the bees or storage of newly foraged honey (Page 30, Figure 5).

TWO DAYS before larval transfer set up 30 hives for royal jelly production as described on Page 25. This involves placing a queen, four empty frames, two pollen and honey frames each and two brood frames in the bottom box below the queen pheromone excluder. Add the second box containing a pollen frame sandwiched between the two frames of queen cells in the middle of the box and on either side one frame of capped/uncapped brood, one frame of pollen and one frame of honey. Add the alternative honey entrance above the second box with a honey super on top in periods of a honey flow. Cover the hive with hardboard hive mat and water-proof lid.

ONE DAY before larval transfer, remove the cover of the lay cage in the nucs so the worker bees can attend the new-laid eggs, allowing the queen to migrate freely throughout the nucleus colony.

7.2.2 Production maintenance

Day 1.   Starting at about 9 am, transfer the plug strips holding the newly-hatched larvae from the nucs to the queen cell frames in the royal jelly hives. Surplus newly-hatched larvae can be discarded. Immediately reset the lay cage from the nuc with clean plugs, and cover. Place the queen inside the cage.

If feeding supplements, place 500g protein patties onto the middle of the pheromone excluder of each royal jelly producing hive.

Day 2.   Remove the cover from the lay cages in the nucs.

Day 3.   Rest day.

Day 4.   At 8 am remove queen cell frames from royal jelly hives and place them in the holding box (Page 54). In the laboratory, remove larvae from cells by tweezers (Page 60, Plate 11) and discard, or save for human consumption. Harvest jelly from the cells by vacuum using the suction apparatus (Page 46). Strain the jelly in the syringe, pack in glass vials or in bulk, and freeze immediately. Remove the used plugs from the queen cell cups and wash them in near-boiling soapy water.

At 1 pm, transfer the plugs of newly hatched larvae into the recently harvest royal jelly hives. In the nucs, reset the plugs in the lay cages using the washed plugs from the morning. Put the queen into the covered cage, and set the lay cage in the nuc.

Day 5.   Remove the cover of the lay cages in the nucs.

Day 6.   Rest day.

Day 7 and beyond.

Harvest the royal jelly, repeat the activities from Day 4.

If supplementary feeding in the morning of every second harvest cycle, add a 500 g protein patty to each hive and 50 g to each nuc.

In the morning of every third harvest cycle, boost royal jelly producing hives with frames of brood, adult bees, pollen and honey as necessary from the surrogate hives. Ensure that foreign queens are not accidentally introduced into the royal jelly producing hives.

Check the frames in the second box for rogue queen cells every six days. Cull any queen cells present.

Replace filled honey supers as required with empty honey supers.

At every harvest check the hives for royal jelly yield and cell acceptance. Hives producing yields below 180 mg/cell and with a cell acceptance below 60% should be replaced with the surrogate hives which have been previously configured for royal jelly production. The surrogate hives need to be at a stage where they can accept newly-hatched larvae in the queen cells. After the third harvest, replace 10% of the lowest performing hives with surrogate hives to maintain and possibly improve yields from the apiary.

7.3 Hobbyist production from one hive

The steps itemised below make recommendations for the production of royal jelly from one hive for personal consumption. A minimum of two colonies are required; one for production, and the other hive to supply supplementary bee resources and newly-hatched larvae when required. Both colonies should contain over 25,000 bees each. Ideally, the surrogate hive should have a recently mated or one-year old queen.

One hive will easily produce 10g of royal jelly every three days. This amount will last one person 20 days consuming 500 mg each day.

7.3.1 Establishment

FOUR DAYS before grafting, place both hives in a sheltered site near your home. Select the hive with the most bees for royal jelly production.

TWO DAYS before grafting, set up the hive for royal jelly production as described on Page 25. Put the queen along with four empty, two pollen and two honey frames in the bottom box. Add the queen pheromone excluder above that and add the second box containing the two frames of queen cells. One or two queen-cell frames should be in the middle of the box. On either side of the queen cells insert two frames of capped and uncapped brood, one frame of pollen and one frame of honey. An additional brood frame can be included to make up 10 frames if only one queen cell frame is used. In periods of a honey flow, add the alternative honey entrance above the second box and place a honey super on top. Cover the hive with a hardboard hive mat and water-proof lid.

7.3.2 Production maintenance

Day 1.   Graft 2-6 hour old hatched larvae from the surrogate hive into the queen cells, and put the frame into the queen-less second box of the royal jelly production hive.

Day 4.   Remove queen cell frames from the hive at a similar time of the day when grafting took place, shaking off the bees as the frame is lifted out of the hive. Place the frame in the holding box.

At home, remove the larvae from cells by tweezers (Plate 11) and discard. Harvest jelly from the cells using a vacuum apparatus (Figure 6), or by using a small spatula. Strain the jelly in the syringe, and pack the royal jelly in glass jars for consumption later.

Graft 2-6 hour old hatched larvae from the surrogate hive into the previously harvested queen cells. Put the frame back in the hive for another cycle.

Day 7   Harvest royal jelly and graft, repeating the cycle as often as required.

At every second harvest, swap empty frames in the second box of the royal jelly producing hive with young brood from the bottom box. Ensure that the queen remains in the bottom box. This interchanging of brood for empty frames in the bottom box will assist in preventing swarming.

Boost the hive with frames of bees, pollen and honey as necessary from the surrogate hive. Check the frames in the second box for rogue queen cells, and if found, cull.

If the colony produces cell yields below 180 mg with about 40 queen cells accepted, then either stop production, or convert the surrogate hive into a royal jelly producer, using the old jelly producing hive as the surrogate hive.

7.4 Time management and yields

Based on the time it takes to complete each necessary task (Table 5), one trained beekeeper can perform all the functions necessary to service a hive in about 15-30 minutes depending on whether the grafting or lay cage regimes are used. About 10 g or more of jelly can be produced from every producing hive every three days.

Table 5: Estimates of labour input (minutes/hive) for each harvest day in managing 30 hives for royal jelly production under grafting or lay cage regimes.

Activity	Grafting (min./hive)	Split plug lay cage (min./hive)
Manipulation of nucs to produce young larvae	3	1
Retrieval & replacement of cell bars	2	2
Selection of grafting frames	2	-
Rogue queen check & frame manipulation	2	2
Grafting or transferring larvae for 60 cells	14	2
Removal and discarding or storage of larvae	3	3
Harvesting and straining royal jelly	5	5
Total time	31	15

Using the grafting technique, one person can service 30 royal jelly production hives, 15 surrogate hives and 10 nucs working a 40 hour week. Fifteen hives are employed for two 8 hour days in succession, with a break on the third day. Using the split plug lay cage system, one person can service 30 royal jelly producing hives, 15 surrogate and five nucs in a 40 hour week. One 8 hour day is worked every three days, with half an hour required every day following harvest to release the queens from the lay cages.

Under both the grafting and lay cage regimes, a minimum of 600 g/week of royal jelly can be expected if production is run for one month. This equates to 700 g/week if production continues for three months because the initial set-up time can be shared over a longer period. These yields are conservative, and can be doubled under favourable conditions. Using the split plug lay cage system, an extra 20 hives could probably be operated for royal jelly production in a 40 hour week. This would entail harvesting the jelly and transferring young larvae to queen cells on the day the nuc queens are released from their cages. Under this scenario, royal jelly yields would increase to 1.0-1.2 kg/week compared to the yields expected by grafting larvae, with no increase in labour costs.

8.   Pest, Disease and Physiological Stress

Pests and diseases are a particular problem in hives producing royal jelly. The regular hive manipulation and harvesting of royal jelly inflicts great stress on the bee colony, making it vulnerable to pests and diseases. Pest and disease symptoms become more pronounced. Colonies are less likely to recover from their effects than if they were a strong colony associated with the minimum disruption under honey production. The regular interchange of brood and hive components, such as grafting tools, queen cells and frames, rapidly spreads even mildly contagious diseases from infected hives to non-infected hives.

In this book it is not possible to cover all the pests and diseases you may encounter. Thus, it is important to be alert to any abnormality in the appearance of brood or bees. Abnormalities usually signify a pest or disease is present, or the colony is suffering from physiological distress. Quickly inspect inside uncapped cells while working the hives, picking off the caps of a few sealed cells to examine the more mature brood.

Become familiar with the appearance of healthy larvae. Healthy larvae have a glistening, pearly-white appearance and distinctive body segmentation

(Plate 3). They lie coiled in a C-shape at the bottom of the cell four days after hatching. Then they lie stretched out along the length of the cell and after the fifth day, the cell is capped with a wax dome. Development of the insect into pre-pupa and pupa continues within the cell. If larvae development is delayed, disease is likely to present. Also remember to become familiar with the appearance of healthy adult worker bees (Plates 1 and Plate 2).

Any bee colonies exhibiting symptoms of pests or diseases should be removed from royal jelly production in the evening (when all bees are in the infected hive) to an isolation area at least 0.5 km away, and replaced with a hive from the surrogate apiary.

The main important pests and diseases that impact on royal jelly production are listed below for quick identification. Remedial action is given for each pest or disease. But because recommendations and legal specification may differ between countries, consult your apicultural advisor first before adopting remedial action.

8.1 American Foul Brood (AFB)

AFB (Plate 14) is caused by a spore-forming bacterium *Paenibacillus larvae larvae*. The bacterium was previously known as *Bacillus larvae*. As little as 10 spores eaten by one-day old larvae with its food can be fatal. Spores germinate in the larval gut, with the resulting vegetative stages of the bacterium forming about 2500 million spores that invade the decomposing larva or pupa. Consequently the disease is highly contagious. The spores are tough and cannot be killed by chemicals or boiling water.

Larvae are most susceptible to AFB infection when they are less than 24 hours old. Millions of spores are required to infect larva more than two days old. Larvae become infected within three days of emergence, but do not die until just before or just after the cells are sealed. Symptoms include scattered patches of capped brood which have been prevented by the disease from emerging, surrounded by cells that either are empty or have unsealed brood in them. Cappings are sunken or concave rather than domed or convex as for healthy larvae, and may become oily-looking and dark. They are often perforated with small holes near the edge of the cells, made by nurse bees as they begin to remove the caps from cells containing the diseased brood.

Infected larvae or pupae show no body segmentation and adhere to the sides of the cell, changing colour from off-white to brown and eventually, after one month the dried corpse becomes a characteristic black scale. The mouth parts of infected pupae may protrude into the middle of the cell when viewed from above after the cap has been carefully removed. When brown, the dead larvae are ropy in texture, and can be drawn out to a 10-30 mm chocolate-brown thread using a twig (Plate 14). This is a characteristic test for identification of the disease, but laboratory tests are required to confirm its presence.

To limit the chance of foraging bees from healthy hives robbing honey from the diseased hive and becoming contaminated themselves, any hives suspected of AFB need to be removed immediately and placed at least 1 km from the apiary. The hive or hives need to be burnt as soon as possible in the evening or during wet weather when there is no likelihood of bees flying. Prior to burning a hive, block the hive entrance and pour a litre of petrol under the lid to kill the colony. After 15 minutes, when the bees have died, burn the hive contents in a ditch. Cover the ashes with soil. Douse hive tools in methylated spirits and flame sterilise. In some countries, boxes, lids and bases in good condition can be stored in a plastic bag until they can be sterilised for 10 minutes in paraffin wax that has been heated to 150°C.

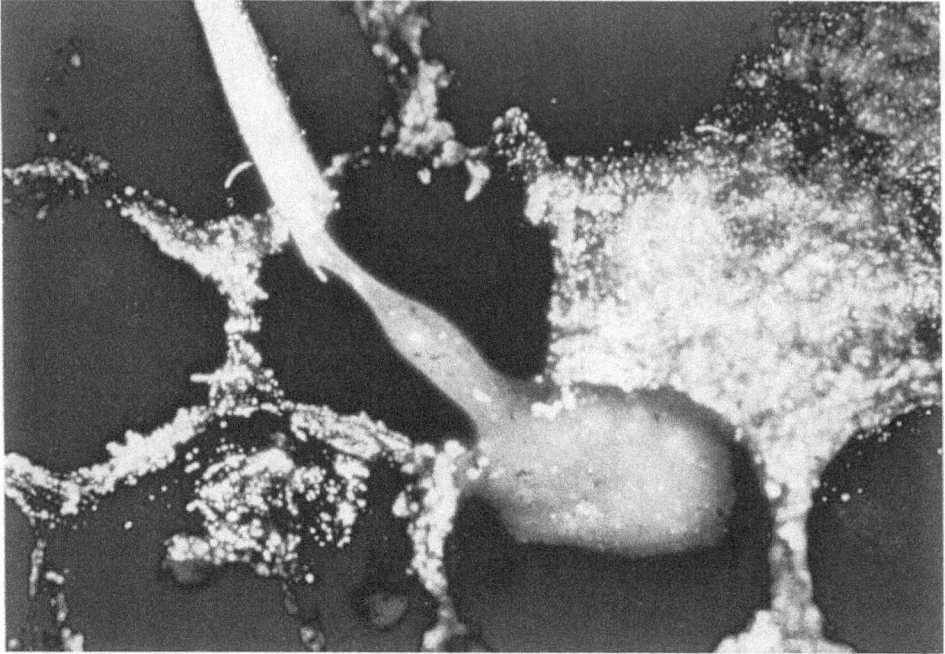

Plate 14: Ropiness test for American Foul Brood

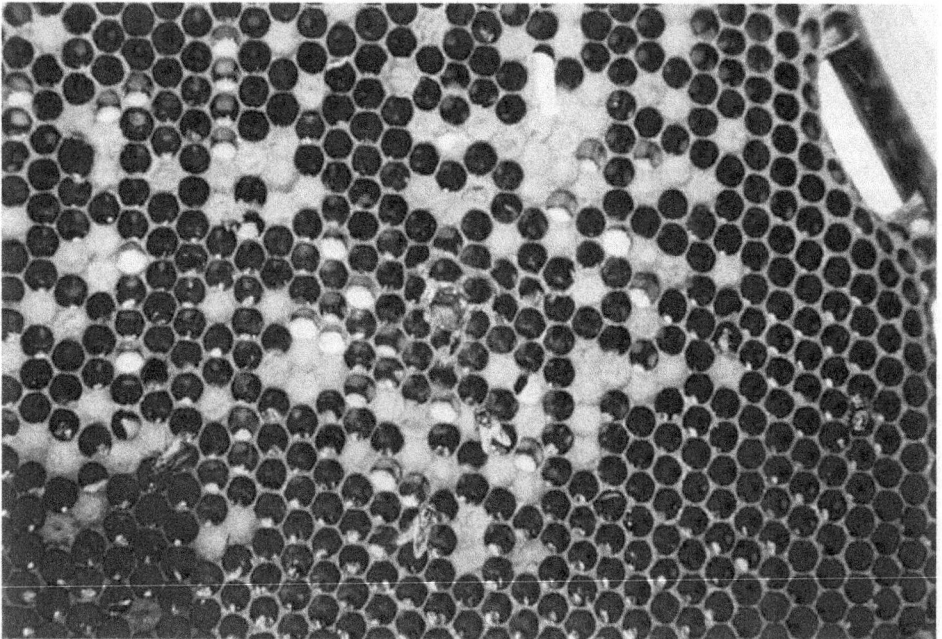

Plate 15: European Foul Brood.

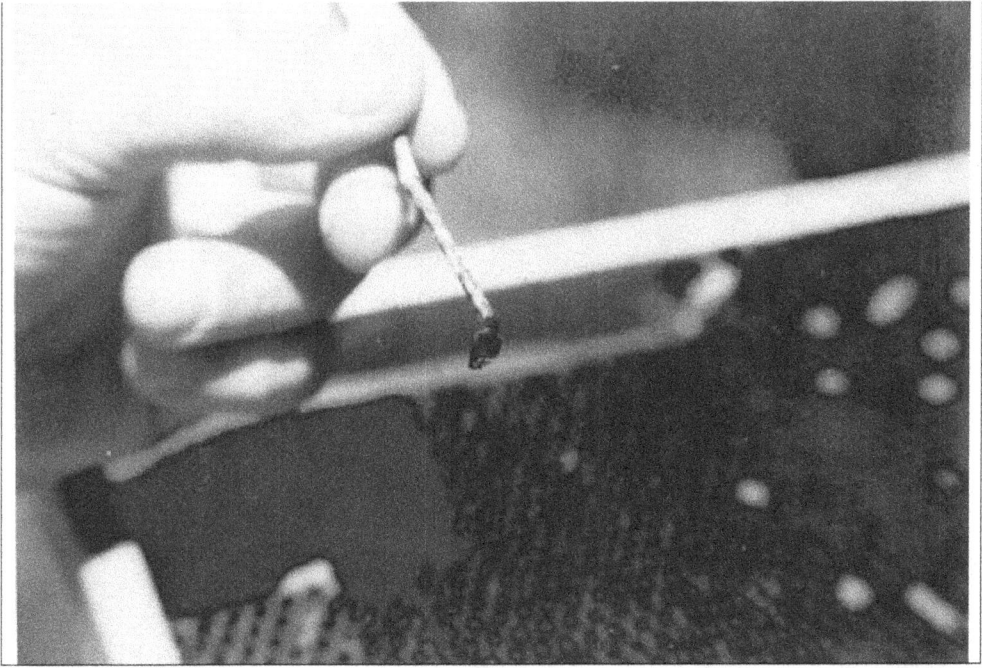

Plate 16: European Foul Brood showing the characteristic brown-black scale.

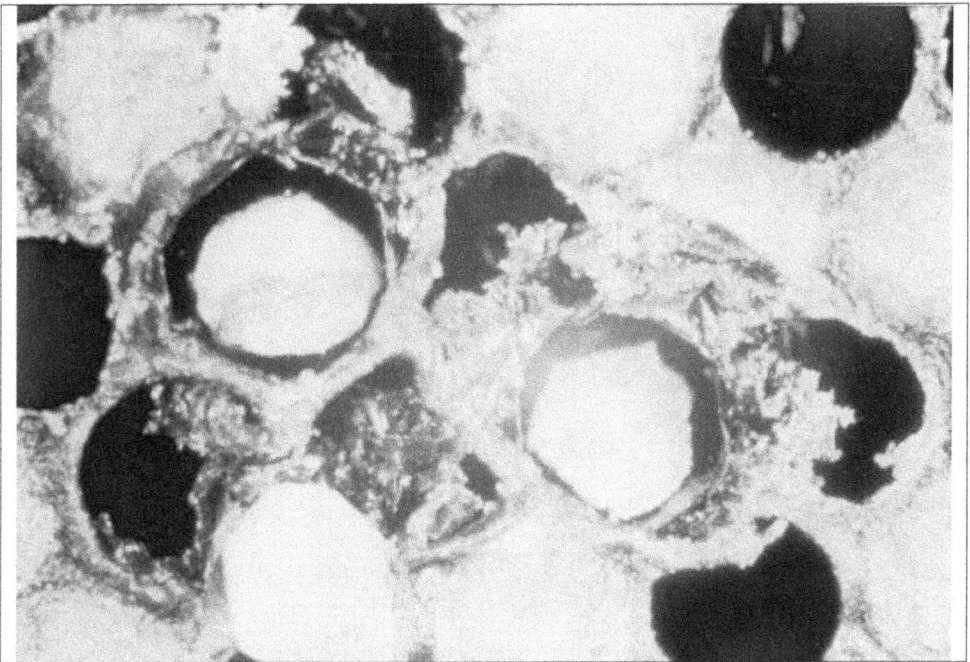

Plate 17: Chalkbrood showing the characteristic white chalk-like mummies.

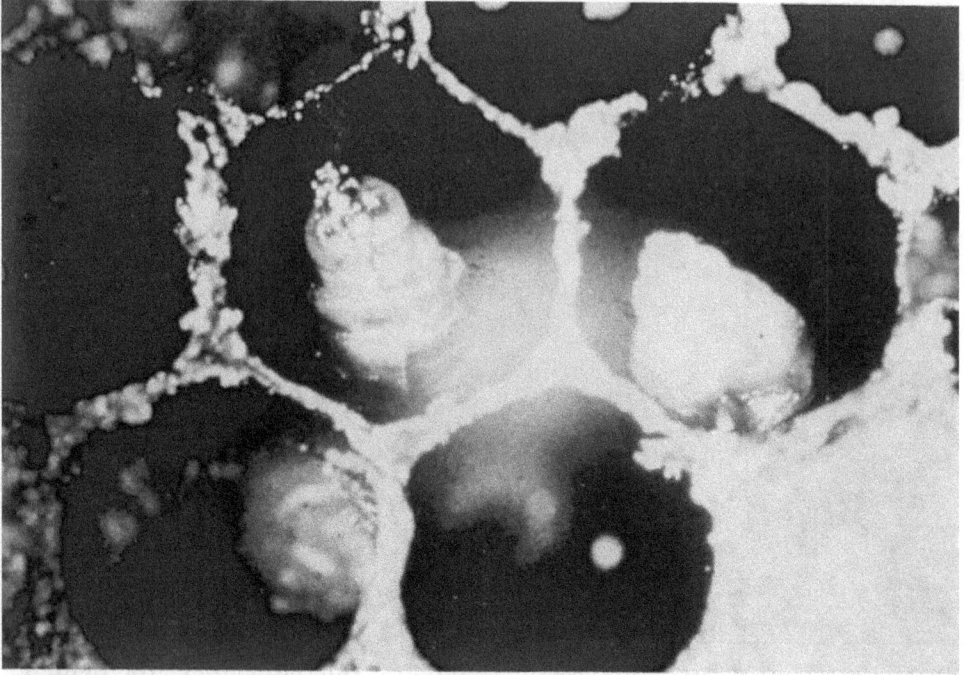

Plate 18: Sacbrood.

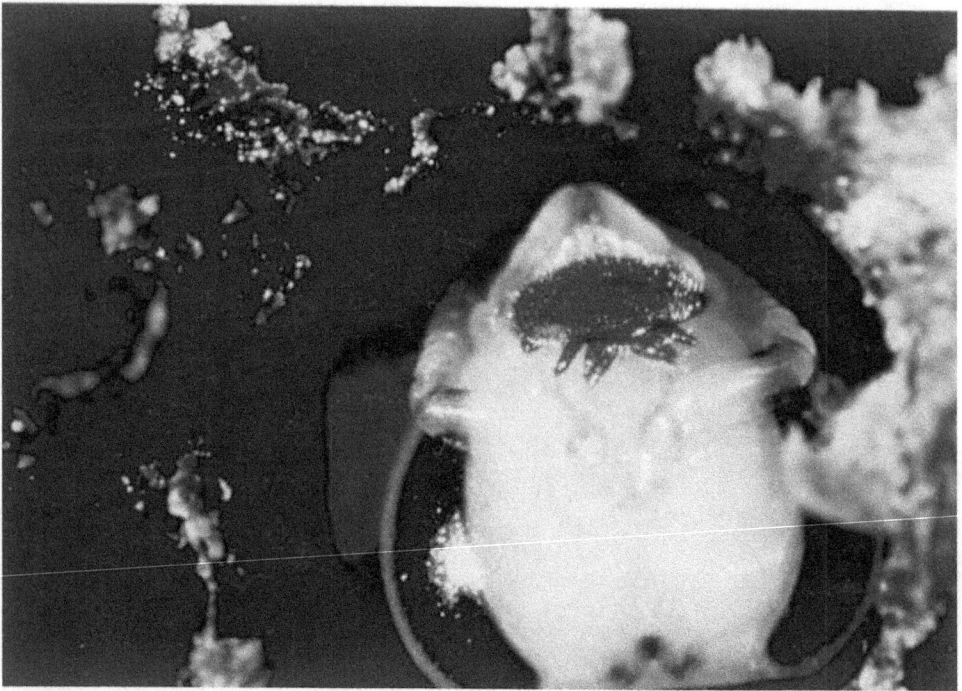

Plate 19: Varroa mite on pupa.

8.2 European Foul Brood (EFB)

EFB is a disease caused by the bacterium *Melissococcus pluton*. It occurs in two day old larvae, usually transmitted by nurse bees performing cleaning and feeding chores. Larvae die at 4-5 days old, and prior to being capped. The disease can over-winter on brood combs and in bee faeces, remaining viable for three years. Infected larvae lose their distinct form, becoming yellow and eventually dark brown. Trachea (breathing tubes) may appear as light lines in the larvae. Prior to dying, the larvae travel up the cell, twisting to appear corkscrewed or half-moon shaped (Plate 15). In medium to heavy infections, brood can appear patchy, with some cappings appearing sunken or perforated, but unlike that of AFB, the infected larvae do not rope out. The thin black scale from desiccated larvae can be easily lifted out of the cell (Plate 16). Infected hives need to be removed from royal jelly production until symptoms have disappeared after natural recovery or after antibiotic treatment.

Oxytetracycline hydrochloride (OTC), is the antibiotic most commonly used to control European Foul Brood in countries where the disease exists. It may be administered to the infected colony (but only after laboratory confirmation of the disease) at least four weeks before the first of harvest of royal jelly. This recommendation is based on trials where OTC residues of 1.7 ppm, which were detected in royal jelly immediately after administration of a commercial formulation, could not be detected after four weeks. Because the break down of OTC is accelerated in the presence of water, feed 1 g OTC active ingredient per hive in a 1:1 sugar-water syrup solution, in preference to the alternative method of feeding the drug with dry sugar. (Check on the legality of using OTC in your country/state).

8.3 Chalkbrood

Chalkbrood is caused by the fungus *Ascosphaera apis*, the spores of which germinate as hyphae in the 3-4 days old larvae and grow inside their gut (Plate 17). Larvae die in the elongated stage in both capped and uncapped cells. The larvae turn vivid white and furry. They become mummified as the cadaver dries. Larvae may become grey to black due to the growth of spores on the surface. Remove the infected hive from royal jelly production and allow time for the bee colony to recover naturally. The cadavers (corpses) are easily removed from the cells by worker bees.

8.4 Sacbrood

Sacbrood (Plate 18) is caused by the sacbrood virus (SBV). Symptoms are similar to AFB; perforated sunken cappings and a patchy brood pattern. Larvae die at the late larval stage after being capped. The head discolours first, followed by the larval body turning yellow, grey and eventually black. The skin resembles a tough, plastic-like sac. The body contents are watery.

Unlike larvae infected with AFB, segmentation is maintained, and dead larvae will not rope. Larvae are easily removed from the cell by worker bees. Remove the hive from royal jelly production and let the bee colony recover without further stress.

8.5 Black queen cell virus

Black queen cell virus causes royal jelly to appear dark brown. The jelly protein coalesces and queen larvae turn black. If these symptoms are present, remove the hive from production. The colony should regain normal health once the stress of royal jelly production is removed.

8.6 Nosema

The levels of *Nosema apis*, a gut protozoan, increase in royal jelly producing hives. Although this disease affects the hypopharyngeal glands of the nurse bees, and therefore possibly their ability to secrete royal jelly, it has not been shown to reduce yields. The disease can be kept at manageable levels by keeping stress levels to a minimum. Reducing the stress on a hive can be achieved by maintaining ventilation under the hive through controlling weeds, or standing the hive on a wooden, stone or concrete pad. Furthermore, ensure at least two frames of pollen and two frames of honey are in the hive at any time during royal jelly production.

In countries where it is permitted, the antibiotic fumagillin can be fed to bees with very high infection levels of *Nosema* (greater than 4 million spores per bee) to reduce the number of spores present in bees to low levels. The hives should be treated in autumn after royal jelly production. Fumagillin can be administered again in spring, but hives need to be treated at least four weeks before royal jelly production to reduce the risk of fumagillin contamination in the jelly. Both times of application are required for effective control. On both occasions an effective dose per hive is 100 mg active ingredient fumagillin mixed with icing sugar. The mix is best applied on to the top bars of the second brood box of each hive. (Check the legality of using this treatment in your country/state).

8.7 Varroa mite

Varroa mite (*Varroa destructor*) is a reddish-brown oval-shaped external parasite 1.0 x 1.6 mm diameter (Plate 19). Mites are most frequently found under the bee's abdomen near its wax glands, and under the cappings of pupae, particularly drone pupae. Emerging bees may appear disfigured, often with misshapen wings and missing legs. The misshapen wings appear to be caused by Deformed Wing Virus, a virus vectored by the mites into bees during feeding. The lifespan of bees infected during their development is significantly reduced. Bees crawling at the hive entrance and patchy brood patterns caused by nurse bees removing infected brood are additional symptoms.

If the colony is heavily infested with varroa, remove the infected hive from royal jelly production and treat the colony with acaracide-impregnated strips such as 'Apistan® or Bayvarol® in late autumn when brood rearing activity is at a minimum and in the following spring during colony build up. The hive can be used for royal jelly production once mite levels have been reduced, immediately following the summer harvest and in the following spring. Royal jelly production can be properly maintained if you follow a well designed Integrated Pest Management (IPM) regime.

8.8 Pest and disease prevention

The following precautions will reduce the risk of contaminating healthy hives and keep stress levels to a minimum.

- Maintain adequate feed levels by feeding protein substitutes or pollen supplements to colonies if production extends beyond five cycles (15 days).

- Allow maximum ventilation around the hives by placing hives on surfaces where plants cannot grow. Appropriate surfaces include concrete, gravel where weeds have been killed with a non-persistent herbicide such as glyphosate, or double-thick cardboard which extends outside the dimensions of the hive by about 300 mm.

- Sterilise the hive tools and grafting tools in an autoclave or pressure cooker for 30 minutes at the end of each daily session involving royal jelly production. Temperatures for effective sterilisation must exceed 120°C or AFB spores will not be killed.

- Do not interchange cell cups between hives. Label the frames containing the cell cups and ensure they go back into the hive they were removed from.

- Manipulate frames over the hives from where they were obtained to prevent honey and wax from dropping outside the hive and being transferred by bees to other colonies. For the same reason, remove wax scrapings from the apiary in plastic bags.

- Limit bee drift to adjacent hives by labelling the entrance to the hives, prudent hive positioning, and painting bee boxes distinctive colours (Page 23).

9.   Packaging and Storage

9.1 Fresh royal jelly

The rate at which royal jelly deteriorates depends on the temperature at which it is stored. If stored at -20°C to -10°C, chemical changes are negligible. Colour and viscosity do not change either. However, slight deterioration of royal jelly refrigerated at 2-4°C occurs within four to six months. Although contents such as amino acids, proteins, reducing sugars, organoleptic properties or pH will not change over one year, a reduction in total acidity and an increase in insoluble matter have been observed. After a year, an increase in amino acid nitrogen content and a decrease in thiamine content have been recorded, but by not more than 9% of the original content.

Using furosine as an indicator of freshness, commercial royal jelly products can contain 37-113 mg furosine/100 g protein, providing evidence of different storage times and conditions. The lower the level of furosine, the fresher the royal jelly. Thus furosine can be used to determine the effects of different storage temperatures on freshness. In one study, furosine content increased from 72 mg/100 g protein to 500 mg/100 g protein after 10 months of storage at room temperature (18-26°C), while it increased to only 100.5 mg/100 g protein when the royal jelly was stored at 4°C. However, nutritional damage, expressed as blocked lysine, was minor or negligible, measured as 11.9 and 2.3% of total lysine in samples stored at room temperature and at 4°C, respectively.

Inhibition of bacteria can remain unchanged. At 25°C deterioration can occur within a month. Some components remain stable over the whole temperature range mentioned. The concentration of 10-Hydroxydecenoic acid, an internationally recognised quality indicator, and microbiological quality, remains stable within -40°C to 25°C.

Therefore, royal jelly can be frozen in a standard domestic freezer at -10°C for up to one year and at 4°C for four months for it still to be considered 'fresh'.

I recommend that the jelly be stored frozen in 10 ml glass containers (e.g. tubes 50 mm long x 20 mm diameter) until about one week prior to intended sale, then displayed in a refrigerator. In this way the jelly retains its light-cream lustre. The vials need to be presented with a distinctive label which includes information on:

1. Price.

2. Net weight.

3. A description of its freshness and that it is natural.

4. 'Best before date' set 16 weeks after the first lot of royal jelly contained in the vial was harvested.

5. Method of storage.

6. Country of origin.

7. Details on how to administer the royal jelly.

8. Recommended daily dose rate.

A plastic spoon, shaped to fit the contour of the inside of the glass, could also be included in the package. The spoon should be dished to hold 250 mg, a standard dose for children. Two spoonfuls would be required for an adult. So that the astringent taste of royal jelly is reduced, the spoon must be designed to easily fit under the tongue to allow for sublingual dissolution.

9.2 International quality standards

The national Royal Jelly Fair Trade Committee's guidelines for quality standards in royal jelly are presented in (Table 6). Fresh royal jelly must not be adulterated with preservatives or any other additives.

Table 6: National Royal Jelly Trade Committee's composition standards on quality of fresh royal jelly for food consumption and medical use.

Components	Detection method	Conc.
Food use		
Moisture content	Drying by heating under vacuum	65.5-68.5%
Crude protein	Kjeldahl	11.0-14.5%
10-hydroxy-2-decenoic acid	Gas chromatography	>1.4%
Acidity (1 N NaOH ml/100g)	Alkaline titration	32-53 ml
Bacterial count	Bacterial cultivation on agar	<500/g
Medical use		
pH		3.5-4.5
Nitrogen	semi-micro Kjeldahl	1.9-2.5%
Sugar		9-13%
Ash		1.5%
Water extract		22-31%
Alcohol extract		14-22%
Heavy metal		<5 ppm
Arsenic		<1 ppm
Tetracycline antibiotics		none
NH_2 radicals		none
Quaternary ammonium		none

9.3 Mixed with honey

This form of packaging allows for easy and convenient storage. It also overcomes the problems caused by the astringent taste of royal jelly and as various enzymatic, antibiotic and aromatic substances in honey may alter the natural composition and structure of royal jelly, possibly affecting its therapeutic properties. Another disadvantage is that the honey-royal jelly mix will probably be swallowed and digested rather than taken sublingually. This will further reduce its potential therapeutic benefits.

To lessen the risk of chemical degradation, royal jelly should be mixed thoroughly into the honey at a concentration of not more than 3.0% as soon as possible before sale. Alternatively, the honey and royal jelly can be sold separately in the proportions to give a 3.0% mixture of honey to royal jelly. Consumers can then add the jelly to the honey immediately prior to consumption, thus also overcoming the problem of chemical degradation..

9.4 Freeze drying

Freeze drying, which is deep freezing and dehydration in a vacuum (lyophilisation), permits long term storage of royal jelly, with no appreciable deterioration of its quality. However, this involves an additional manufacturing expense and is outside the scope of this book.

10. Bibliography

Composition

Benfenati, L.; Sabatini, A.G.; Nanetti, A. 1986: Mineral composition of royal jelly. *Apicoltura, Italy (No.2):* 129-143.

Cerna, J., 1963: Vitamins in royal jelly. *Potravin. Storjir. Chlad. Tech No. 139:* 192-196.

Glinski, Z.; Rzedzicki, J. 1975: Analysis of royal jelly of honey bees. I. Amino acid composition of hydrolysates of fresh and lyophilised royal jelly. *Polskie Archiwum Weterynarynje 18(3):* 411-418(B).

Howe, S.R.; Dimick, P.S. and Benton, A.W.; 1985: Composition of freshly harvested and commercial royal jelly. *Journal of Agricultural Research 24(1):* 52-61.

Ishiguro, I.; Naito, J.; Shinowara, R. & Watanable, M. 1963(b): Nutritional studies on royal jelly Part VII. Effects on royal jelly on the system of internal section. *Acta Sch. med. Gifu 13:* 21-27.

Kramer, K.J.; Childs, C.N.; Speirs, R.D.; Jacobs, R.M. 1982: Purification of insulin-like peptides from insect haemolymph and royal jelly. *Insect Biochemistry 12(1):* 91-98(B).

Marconi, E.; Caboni, M. F.; Messia, M. C.; Panfili, G. 2002: Furosine: a suitable marker for assessing the freshness of royal jelly. *Journal of Agricultural and Food Chemistry 50 (10):* 2825-2829 .

Matsuka, M. 1993: Content of benzoic acid in royal jelly and propolis. *Honeybee Sci 14(2):* 79-80.

Nakamura, T. 1986: Quality standards of royal jelly for medical use. *Proceedings XXXth International Congress of Apiculture, Nagoya, 1985, Apimondia:* 462-464.

Takenaka, T. 1984: Studies on proteins and carboxylic acid in royal jelly. *Bulletin of the Faculty of Agriculture, Tamagaura University.* (No. 24): 101-149.

Takenaka, T.; Echigo, T. 1980: Chemical composition of royal jelly. *Bulletin of the Faculty of Agriculture, Tamagaura University,* (No. 77.20): 71-78.

Vittek, J. and Slominay, B.L. 1984: Testosterone in royal jelly. *Experientia* 40 pp 104-106.

Management

van Toor, R.F. and Littlejohn, R.P, 1994. Evaluation of hive management techniques in production of royal jelly by honeybees (*Apis Mellifera*) in New Zealand. *Journal of Apicultural Research 33(3):* 160-166.

Zaytoon, A.; Matsuka, M.; Saski, M. 1988. Feeding efficiency of pollen substitutes in a honeybee colony; effect of feeding site on royal jelly and queen production. *Applied Entomology and Zoology 23(4):* 481-487

Medical

Bincoletto, C.; Eberlin, S.; Figueiredo, C. A. V.; Luengo, M. B.; Queiroz, M. L. S. 2005: Effects produced by royal jelly on haematopoiesis: relation with host resistance against Ehrlich ascites tumour challenge. *International Immunopharmacology 5 (4):* 679-688.

Bonomi, A. 2003. Royal jelly in heavy pig feeding. *Rivista di Suinicoltura 44 (8):* 55-61.

Derevici, A.; Petrecu, A. 1965(a): The effect of the water-soluble extract of royal jelly on certain viruses. *Vop. Virusol. 6(5):* 6111-614.

Derevici, A.; Petrecu, A. 1965(b): Action of the water soluble fraction of royal jelly (FAPA) against viruses. Note 1. Experiments on the influenza virus. *Commun. Acad. RPR 9*(12): 1337-1342.

Donadieu, Y,. 1983: Royal jelly in natural therapeutics. Paris, France; Maloine Editeur S.A. Edition 6, 56 pp.

Iizuka, H. and Koyama, Y. 1964: Studies on royal jelly (Part III). Antibiotic actions of royal jelly. *Japanese Society Fd Nutr.* 17(3) 203-207.

Inoue, S.; Koya-Miyata, S.; Ushio, S.; Iwaki, K.; Ikeda, M.; Kurimoto, M. 2006: Royal Jelly prolongs the life span of C3H/HeJ mice: correlation with reduced DNA damage. *Experimental Gerontology 38 (9):* 965-969.

Ishiguro, U.; Naito, J. and Harada, J. 1963(a): Nutritional studies on royal jelly (Part V). Growth effect on rat by royal jelly. *Acta Sch. med. Gifu* 13: 17-21.

Kohguchi, M.; Inoue, S.; Ushio, S.; Iwaki, K.; Ikeda, M.; Kurimoto, M., 2004: Effect of royal jelly diet on the testicular function of hamsters. *Food Science and Technology Research 10 (4):* 420-423.

Kramer, K.J.; Tager, H.S.; Childs, C.N.; Speirs, R.D. 1977: Insulin-like hypoglycaemic and immunological activities in honeybee royal jelly. *Journal of Insect Physiology* 23(2): 293-295.

Kridli R. T. and Al-Khetib S. S. 2006: Reproductive responses in ewes treated with eCG or increasing doses of royal jelly. *Animal Reproduction Science 92 (1/2):* 75-85.

Kushima, S. 1986: On the medical efficacy of royal jelly. *International Congress of Apiculture Proceedings of the 30th, Nagoya. 30:* 448-452.

Lakin, A. 1993: Royal jelly and its efficacy. *International Journal of Alternative & Complementary Medicine 11(10):* 19-22.

Mishima, S.; Suzuki, K. M.; Isohama, Y.; Kuratsu, N.; Araki, Y.; Inoue, M.; Miyata, T. 2005: Royal jelly has estrogenic effects in vitro and in vivo. *Journal of Ethnopharmacology 101 (1/3):* 215-220.

Munstedt, K. and von Georgi, R. 2003: Royal jelly - a miraculous product from the bee hive? *American Bee Journal 143 (8):* 647-650.

Rice, W.B. & Lu, F.S. 1964: A quantitative study of acetylcholine in royal jelly preparations. *Canadian Pharmacy Journal* 97(3): 34-35.

Stein,E. 1986: Royal jelly, a guide to nature's richest health food. Wellingborough, U.K.; Thorsons Publishers.

Townsend, G.F.; Joseph, F.M.; Hazlett, B. 1959: 2. Activity of 10-Hydroxydecenoic Acid from royal jelly against experimental leukaemia and ascitic tumours. *Nature Vol. 183 No. 4670.*

Vittek, J.; Kresanek, J.; Ditte, L.; Svrec, D. 1962: Royal jelly (RJ) - gelee royale - of domestic origin investigated from pharmaceutical point of view and possibilities of its therapeutical application in stomatology. *Acta Fac. Pharm. bohemsl.* 7: 93-122.

Yatsunami, K.; Echigo, T. 1984: Antibacterial activity of honey and royal jelly. *Honeybee Science 5(3):* 125-130.

Young, T. 1977: Studies on royal jelly and abnormal cholesterol and triglycerides. *American Bee Journal:* pp 36-38.

Pesticide residues

Matsuka, M.; Nakamura, J. 1990. Oxytetracycline residues in honey and royal jelly. *Journal of Apicultural Research, 29(2)*: 112-117.

Storage

Lee E.L.; Chu L.K.; Li C.F.; Hsu E.L. 1988: Changes of the components of royal jelly stored at different temperatures. *Shih-Pin-K'o-Hsueh, Taipei 15(1): 81-90.*

Takenaka, T.; Yatsunami, K; Echigo, T. 1986: Changes in quality of royal jelly during storage. *Nippon-Shokuhin-Kogyo-gakkaishi 33(1): 1-7.*

Fik, M; Firek, B; Leszynska-Fik, A; Macura, R; Surowka, K. 1992: Effect of refrigerated storage on the quality of royal jelly. *Pszczelinicze-Zeszyty-Naukowe 36:* 91-101.

11. Acknowledgements

Ruth Butler gleaned relevant information, and provided encouragement to complete this manuscript. Noel Johnson and MAF Quality Management supplied some photographs. Noel and Annette Johnson provided information on the split peg transfer system, and the initial impetus to write this book. Northern Bee Books provided permission to use the quote at the front of this book.

Thanks also to Kathy Ballantyne for all her support whilst I ran trials on royal jelly production. And to Ben and Dot Rawnsley for their encouragement and supply of bee hives from which I gained much of the experience which formed the basis for writing this book.

11.1 Photograph credits

Plate 1: Healthy adult drone (left), worker (top right) and queen (bottom right).   H van Puffelen

Plate 2: Formation of natural queen cells.   R van Toor

Plate 3: An egg (top left), and healthy worker larvae (top row) and pupae (bottom row) at different development stage.   H van Puffelen

Plate 4: An ideal site for royal jelly production.   B Barratt

Plate 5: A full-depth frame covered by approximately 3,500 bees.   N Johnson

Plate 6: Selection of 2-6 hour old larvae (inset) suitable for grafting, inclining the frame so the cells point directly upwards.   N Johnson, B Barratt

Plate 7: Retractable grafting quill used to transfer newly hatched larvae to the queen cells.   B Barratt

Plate 8: Workers drawing out comb in a lay cage. A queen being put into the cage (inset).   N Johnson

Plate 9: Transferring the plugs containing young larvae (inset) from the lay cage to the cell bars using the 'EZI Queen larvae transfer system'.   N Johnson

Plate 10: Wax drawn out at the entrance of accept cells.   N Johnson

Plate 11: Discarding larvae in preparation for harvesting.   B Barratt

Plate 12: Harvesting royal jelly with a suction apparatus.   B Barratt

Plate 13: 10 g royal jelly harvested from 44 queen cells.   B Barratt

Plate 14: Ropiness test for American foul brood   MAF Qual NZ

Plate 15: European Foul Brood.   R Price

Plate 16: European foulbrood showing the characteristic brown-black scale.   R Price

Plate 17: Chalkbrood showing the characteristic white chalk-like mummies.   MAF Qual NZ

Plate 18: Sacbrood.   MAF Qual NZ

Plate 19: Varroa mite on pupae.   MAF Qual NZ

Cover jacket   N Johnson